Human Factors Considerations in the Design and Evaluation of Electronic Flight Bags (EFBs)

Version 2

DOT/FAA/AR-03/67
DOT-VNTSC-FAA-03-07

Office of Aviation Research
Washington, DC 20591

Divya C. Chandra
Michelle Yeh
U.S. Department of Transportation
Research and Special Programs Administration
John A. Volpe National Transportation Systems Center
Cambridge, MA 02142

Vic Riley
User Interaction Research and Design, Inc.
Point Roberts, WA 98281

Susan J. Mangold
Battelle Memorial Institute
Columbus, OH 43201

September 2003

This document is available to the public through the National Technical Information Service, Springfield, Virginia, 22161

Notice

This document is disseminated under the sponsorship of the Department of Transportation in the interest of information exchange. The United States Government assumes no liability for its contents or use thereof.

Notice

The United States Government does not endorse products or manufacturers. Trade or manufacturers' names appear herein solely because they are considered essential to the objective of this report.

REPORT DOCUMENTATION PAGE

Form Approved
OMB No. 0704-0188

Public reporting burden for this collection of information is estimated to average 1 hour per response, including the time for reviewing instructions, searching existing data sources, gathering and maintaining the data needed, and completing and reviewing the collection of information. Send comments regarding this burden estimate or any other aspect of this collection of information, including suggestions for reducing this burden, to Washington Headquarters Services, Directorate for Information Operations and Reports, 1215 Jefferson Davis Highway, Suite 1204, Arlington, VA 22202-4302, and to the Office of Management and Budget, Paperwork Reduction Project (0704-0188), Washington, DC 20503.

1. AGENCY USE ONLY (Leave blank)	2. REPORT DATE September 2003	3. REPORT TYPE AND DATES COVERED Final Report, September 2003
4. TITLE AND SUBTITLE Human Factors Considerations in the Design and Evaluation of Electronic Flight Bags Version 2		5. FUNDING NUMBERS A3720/FA3E2
6. AUTHOR(S) Divya C. Chandra, Michelle Yeh, Vic Riley *, Susan J. Mangold**,		
7. PERFORMING ORGANIZATION NAME(S) AND ADDRESS(ES) U.S. Department of Transportation John A. Volpe National Transportation Systems Center Research and Special Programs Administration Cambridge, MA 02142-1093		8. PERFORMING ORGANIZATION REPORT NUMBER DOT-VNTSC-FAA-03-07
9. SPONSORING/MONITORING AGENCY NAME(S) AND ADDRESS(ES) U.S. Department of Transportation Federal Aviation Administration Office of Aviation Research, Human Factors Research and Engineering Division 800 Independence Avenue, SW Washington, D.C. 20591 Program Manager: Dr. Tom McCloy		10. SPONSORING/MONITORING AGENCY REPORT NUMBER DOT/FAA/AR-03/67
11. SUPPLEMENTARY NOTES * User Interaction Research and Design, Point Roberts, WA ** Battelle Memorial Institute, Columbus, OH		
12a. DISTRIBUTION/AVAILABILITY STATEMENT This document is available to the public through the National Technical Information Service, Springfield, VA 22161		12b. DISTRIBUTION CODE

13. ABSTRACT (Maximum 200 words)

Electronic Flight Bags (EFBs) are coming into the flight deck, bringing along with them a wide range of human factors considerations. In order to understand and assess the full impact of an EFB, designers and evaluators require an understanding of how the device will function and be used by crews, how the device will interact with other flight deck equipment, and how training and operating procedures will be affected. The purpose of this report is to identify and prioritize guidance on these topics so that designers and evaluators can make informed choices. Much of the guidance in this document is general and applies to any EFB system, regardless of the applications that are supported. Application-specific guidance is also provided for electronic documents, electronic checklists, flight performance calculations, and electronic charts. In addition, information on the rapidly changing and growing market of EFB products is provided in Appendix A, and a summary of high priority guidance for equipment evaluations is included in Appendix B. This document supersedes the earlier Version 1 report (DOT-VNTSC-FAA-00-22), which is referenced in the Federal Aviation Administration Advisory Circular on EFBs, AC 120-76A.

14. SUBJECT TERMS Electronic Flight Bag, EFB, flight deck technology, electronic documents, electronic checklists, ECL, flight performance calculations, electronic charts, usability, avionics, human factors, design, evaluation			15. NUMBER OF PAGES 206
			16. PRICE CODE
17. SECURITY CLASSIFICATION OF REPORT Unclassified	18. SECURITY CLASSIFICATION OF THIS PAGE Unclassified	19. SECURITY CLASSIFICATION OF ABSTRACT Unclassified	20. LIMITATION OF ABSTRACT

NSN 7540-01-280-5500

Standard Form 298 (Rev. 2-89)
Prescribed by ANSI Std. 239-18
298-102

METRIC/ENGLISH CONVERSION FACTORS

ENGLISH TO METRIC	METRIC TO ENGLISH
LENGTH (APPROXIMATE)	**LENGTH** (APPROXIMATE)
1 inch (in) = 2.5 centimeters (cm) 1 foot (ft) = 30 centimeters (cm) 1 yard (yd) = 0.9 meter (m) 1 mile (mi) = 1.6 kilometers (km)	1 millimeter (mm) = 0.04 inch (in) 1 centimeter (cm) = 0.4 inch (in) 1 meter (m) = 3.3 feet (ft) 1 meter (m) = 1.1 yards (yd) 1 kilometer (km) = 0.6 mile (mi)
AREA (APPROXIMATE)	**AREA** (APPROXIMATE)
1 square inch (sq in, in^2) = 6.5 square centimeters (cm^2) 1 square foot (sq ft, ft^2) = 0.09 square meter (m^2) 1 square yard (sq yd, yd^2) = 0.8 square meter (m^2) 1 square mile (sq mi, mi^2) = 2.6 square kilometers (km^2) 1 acre = 0.4 hectare (he) = 4,000 square meters (m^2)	1 square centimeter (cm^2) = 0.16 square inch (sq in, in^2) 1 square meter (m^2) = 1.2 square yards (sq yd, yd^2) 1 square kilometer (km^2) = 0.4 square mile (sq mi, mi^2) 10,000 square meters (m^2) = 1 hectare (ha) = 2.5 acres
MASS - WEIGHT (APPROXIMATE)	**MASS - WEIGHT** (APPROXIMATE)
1 ounce (oz) = 28 grams (gm) 1 pound (lb) = 0.45 kilogram (kg) 1 short ton = 2,000 pounds (lb) = 0.9 tonne (t)	1 gram (gm) = 0.036 ounce (oz) 1 kilogram (kg) = 2.2 pounds (lb) 1 tonne (t) = 1,000 kilograms (kg) = 1.1 short tons
VOLUME (APPROXIMATE)	**VOLUME** (APPROXIMATE)
1 teaspoon (tsp) = 5 milliliters (ml) 1 tablespoon (tbsp) = 15 milliliters (ml) 1 fluid ounce (fl oz) = 30 milliliters (ml) 1 cup (c) = 0.24 liter (l) 1 pint (pt) = 0.47 liter (l) 1 quart (qt) = 0.96 liter (l) 1 gallon (gal) = 3.8 liters (l) 1 cubic foot (cu ft, ft^3) = 0.03 cubic meter (m^3) 1 cubic yard (cu yd, yd^3) = 0.76 cubic meter (m^3)	1 milliliter (ml) = 0.03 fluid ounce (fl oz) 1 liter (l) = 2.1 pints (pt) 1 liter (l) = 1.06 quarts (qt) 1 liter (l) = 0.26 gallon (gal) 1 cubic meter (m^3) = 36 cubic feet (cu ft, ft^3) 1 cubic meter (m^3) = 1.3 cubic yards (cu yd, yd^3)
TEMPERATURE (EXACT)	**TEMPERATURE** (EXACT)
[(x-32)(5/9)] °F = y °C	[(9/5) y + 32] °C = x °F

QUICK INCH - CENTIMETER LENGTH CONVERSION

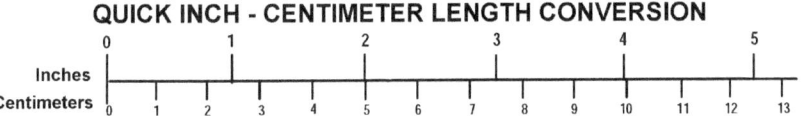

QUICK FAHRENHEIT - CELSIUS TEMPERATURE CONVERSION

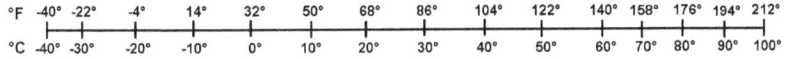

For more exact and or other conversion factors, see NIST Miscellaneous Publication 286, Units of Weights and Measures. Price $2.50 SD Catalog No. C13 10286 Updated 6/17/98

Trademark Notices

The following information was gathered from speaking with company representatives and by searching the database on the United States Patent and Trademark Office website. Other product and company names mentioned herein may be the trademarks of their respective owners. We cannot guarantee the accuracy of the contents of this document. We disclaim liability for errors, omissions or future changes.

ACARS is a registered trademark of ARINC Inc.

Adobe, Adobe Acrobat, and Acrobat Reader are the registered trademarks of Adobe Systems, Inc.

Advanced Data Research, FG-3600, FG-5000, and FG-8000 are trademarks of Advanced Data Research, Inc.

Airbus is the registered trademark of Deutche Airbus GMBH.

Aircraft Data Fusion and *x*EFB are the trademarks of Duane Edelman.

Anywhere Map and ControlVision are the registered trademarks of Control Vision Corporation.

Avionics Magazine is the registered trademark of Montrose Publishing, Inc.

AvVantage is the registered trademark of Spirent Systems Wichita, Inc.

Boeing, Boeing Business Jets, Connexion, and 777 are the registered trademarks of Boeing Company.

CMC Electronics is the registered trademark of CMC Electronics Inc.

C-Map/Aviation, AvMap, EKP-IIIC, and EKP-IIIC Pro are the registered trademarks of C-MAP/USA, Inc.

Cessna and Citation are the registered trademarks of Cessna Aircraft Company.

Challenger and Global Express are the registered trademarks of Bombardier Inc.

Chartrax is a trademark of FlitePath Corporation.

Compaq and IPAQ are the registered trademarks of Compaq Information Technologies Group, L.P. CPQ Holdings, Inc.

Continental is the registered trademark of Continental Airlines, Inc.

DHL is the registered trademark of DHL Corporation.

eflightpad is a trademark of **eflight**systems, LLC.

Embraer is the registered trademark of Embraer - Empresa Brasileira de Aeronautica S.A.

eMonocle is the registered trademark of ION Systems, Inc.

FedEx is the registered trademark of Federal Express Corporation.

Flight International is the registered trademark of Reed Business Publishing Limited.

Flightman and eTechLog are the registered trademarks of Aircraft Management Technologies.

Flight Deck Resources, SkyTab, and FDMS are the trademarks of Flight Deck Resources.

Flight Prep or FlightPrep.com are registered trademarks of Stenbock & Everson Inc.

Flitesoft and Vista are trademarks of RMS Technology, Inc.

FlyTimer is the registered trademark of FlyTimer Corporation.

Fujitsu is the registered trademark of Fujitsu Limited.

Gulfstream is the registered trademark of Gulfstream Aerospace Corporation.

HP is the registered trademark of Hewlett-Packard Development Company, L.P. HPQ Holdings, LLC, a Delaware corporation

Honeywell is the registered trademark of Honeywell International Inc.

Jeppesen, FliteDeck, FliteMap, FliteStar, and JeppView are the registered trademarks of Jeppesen Sanderson, Inc.

JeppView images were provided and reproduced with the permission of Jeppesen Sanderson, Inc. They are not to be used for navigation and have been reduced for illustrative purposes.

Jet Aviation is the registered trademark of Jet Aviation International, Inc.

King Air is the registered trademark of Beech Aircraft Corporation.

Linux is the registered trademark of Linus Torvalds.

Microsoft, Microsoft Windows, Windows 2000, Windows operating system, and Windows XP are the registered trademark of Microsoft Corporation.

Nav 2000 is a registered trademark of JP Instruments

Paperless Cockpit is a trademark of Paperless Cockpit.

NavAero, t-Bag, and t-Pad are the registered trademarks of Navaero Inc.

Qantas is the registered trademark of Qantas Airways Limited.

Panasonic and Toughbook are the registered trademarks of Matsushita Electric Industrial Co., Ltd.

Pentium is the registered trademark of Intel Corporation or its subsidiaries in the United States and other countries.

R-A-M is the registered trademark of National Products, Inc.

Rockwell Collins is the registered trademark of Rockwell International Corporation.

SAMM is a registered trademark of GSCS, Inc.

Stylistic is a trademark of Fujitsu Computer Systems.

Symbian OS is the registered trademark of Symbian Limited.

Teledyne is the registered trademark of Teledyne Technologies, Inc.

UNIX is the registered trademark of The Open Group.

Universal is the registered trademark of Universal.

WxWorx on Wings is the registered trademark of Baron Services, Inc.

WSI is the registered trademark of WSI Corporation.

WalkAbout Computers and Hammerhead 3 are the registered trademarks of Walkabout Computers, Inc.

Preface

There is currently great interest and activity towards developing small electronic information management devices for use by flight crew in performing flight-related tasks. These devices, referred to as "Electronic Flight Bags" (EFBs), aid pilots and aircraft operators in conducting flights more efficiently and safely. EFBs were originally seen as a repository for electronic documents such as checklists, operating manuals, and navigation publications, but now they are seen as multi-function devices that can support an array of applications beyond those of a traditional flight bag, including cabin surveillance, surface moving map, electronic messaging, and display of live weather. Some EFBs will even be fully installed systems with multiple functions.

On March 17, 2003, the Federal Aviation Administration (FAA) issued an Advisory Circular (AC) on EFBs (AC 120-76A). The AC was written jointly by the Offices of Flight Standards and Aircraft Certification. It provides a relatively streamlined approval and certification process for EFBs that meet specific criteria related to installation and functionality. In part due to the EFB AC, the EFB industry is flourishing and many new products are coming to market for all types of operations.

One section of AC 120-76A is devoted to human factors considerations. Much of the content in this section was culled from an earlier version (DOT-VNTSC-FAA-00-22) of the current report. This new Version 2 document supersedes the earlier Version 1 report. This updated Version 2 document contains information for FAA evaluators, system designers/manufacturers and operators about the many human factors considerations that may be associated with EFB that need to be addressed during the design and evaluation of the EFB.

This report was prepared by the Operator Performance and Safety Analysis Division of the Office of Research and Analysis at the Volpe Center. It was completed under the Division's Flight Deck Technology Human Factors program under the sponsorship of the FAA's Human Factors Research and Engineering Division. Dr. Tom McCloy served as the FAA program manager. The participation of Susan Mangold and Vic Riley in this effort was funded by the Volpe Center (DTRS57-99-D-0030 RA 0305).

The authors would especially like to thank Bill LeRoy, Jim Hartman, members of the Air Transport Association Digital Display Working Group, and the many other industry experts and FAA staff who reviewed the report and provided valuable feedback. This final report is the result of all of their efforts as well as ours.

Feedback on this document can be sent to Divya Chandra (Chandra@volpe.dot.gov) or Michelle Yeh (Yeh@volpe.dot.gov).

Further information on this research effort can be found at http://www.volpe.dot.gov/opsad/efb.

Table of Contents

1 Introduction 1
   1.1   Background 1
   1.2   How to Use this Document 2
2 General EFB System 5
   2.1   General Considerations 6

2.1.1	Workload	6
2.1.2	Using EFBs During High Workload Phases of Flight	8
2.1.3	Compatibility and Consistency with Flight Deck Systems and Other Flight Information	10
2.1.4	Use of EFB with Other Flight Deck Systems	12
2.1.5	Legibility—Lighting Issues	13
2.1.6	System Shutdown	15
2.1.7	Failure Modes	16

   2.2   Physical Considerations 18

2.2.1	Design and Placement of Structural Cradle	18
2.2.2	Stowage Area	20
2.2.3	Use of Unsecured EFBs	21
2.2.4	Kneeboard EFBs	22

   2.3   Training/Procedures Considerations 24

2.3.1	Part 121, Part 125, and Part 135 Operations EFB Policy	24
2.3.2	EFB Documentation for Part 121, Part 125, and Part 135 Operations	25
2.3.3	EFB Documentation for Part 91 Operators	26
2.3.4	Initial EFB Training for Part 121 and Part 135 Operators	27
2.3.5	Evaluating EFB Proficiency for Part 121 and Part 135 Operators	29
2.3.6	Fidelity of EFB Training Device	30
2.3.7	User Feedback	31

   2.4   Software Considerations 32

2.4.1	User Interface—General Design	32
2.4.2	Application Compatibility and Style Guides	34
2.4.3	General Use of Colors	36
2.4.4	Graphical Icons	39
2.4.5	Multi-Tasking	41
2.4.6	Responsiveness	43
2.4.7	Anchor Locations	45
2.4.8	Alerts and Reminders	46
2.4.9	Display of System Status	48
2.4.10	Legibility of Text—Characters	49
2.4.11	Legibility of Text—Typeface Size and Width	50

	2.4.12	Legibility of Text—Spacing for Readability	52
	2.4.13	Non-Text Display Elements	53
	2.4.14	Supplemental Audio	55
	2.4.15	Ensuring Integrity of EFB Data	56
	2.4.16	Updating EFB Data	58
	2.4.17	Crew Confirmation of EFB Software/Database Approval	59
	2.4.18	Links to Related Material	60
	2.4.19	User-Interface Customization	62
2.5		Hardware Considerations	64
	2.5.1	Pointing and Cursor Control Devices	64
	2.5.2	Hardware Controls	66
	2.5.3	Display	69
	2.5.4	Accessibility of Hardware Components	71
	2.5.5	Keyboards	72
3		Electronic Documents	73
3.1		Background	73
	3.1.1	Type of Documents Addressed	73
	3.1.2	Features of Electronic Documents	73
3.2		General	76
	3.2.1	Consistency of Information Structure	76
	3.2.2	Training Needs	77
3.3		Layout/Appearance	78
	3.3.1	Visual Layout and Structure	78
	3.3.2	Minimum Display Area and Resolution	79
	3.3.3	Off-Screen Text	80
	3.3.4	Active Regions	81
	3.3.5	Display of High Priority Information	82
	3.3.6	Figures	83
	3.3.7	Tables	85
3.4		Navigation and Searching	86
	3.4.1	Moving to Specific Locations	86
	3.4.2	Managing Multiple Open Documents	87
	3.4.3	Searching	88
3.5		Options	89
	3.5.1	Printing	89
	3.5.2	Animation	90
	3.5.3	Making Notes	91
	3.5.4	Decision Aid/Automatic Call-up of Data	92
4		Electronic Checklist Systems	93
4.1		Background	93

	4.1.1	Research on Paper and Electronic Checklists	93
	4.1.2	Features of ECL Systems	93
4.2		General	96
	4.2.1	Checklists Supported by the ECL System	96
	4.2.2	Information and Visual Layout/Structure of Electronic Checklists	97
	4.2.3	Checklist Data and its Modification	98
4.3		Interacting with Checklists	100
	4.3.1	Accessing Checklists	100
	4.3.2	Managing Checklists	102
	4.3.3	Managing Non-Normal Checklists	104
	4.3.4	Lengthy Checklists	105
	4.3.5	Closing or Completing a Checklist	107
	4.3.6	Closing All Checklists	108
4.4		Interacting with Checklist Items	109
	4.4.1	Indicating the Active Item	109
	4.4.2	Displaying Item Status	110
	4.4.3	Moving Between Items Within a Checklist	111
	4.4.4	Specifying Completion of Item	113
4.5		Options	115
	4.5.1	Links Between Checklist Items and Related Information	115
	4.5.2	Links to Calculated Values	117
	4.5.3	Task Reminders	118
	4.5.4	Checklist Branching	119
5 Flight Performance Calculations			121
	5.1.1	Default Values	122
	5.1.2	Data-entry Screening and Error Messages	123
	5.1.3	Support Information for Performance Data Entry	124
	5.1.4	When and How to Do Performance Calculations	125
	5.1.5	Modifying Performance Calculations	126
	5.1.6	Use of Performance Calculation Output	127
	5.1.7	Coordination Between Flight Crews and Ground Dispatch Personnel	128
	5.1.8	Aircraft Performance Documentation	129
6 Electronic Charts			131
6.1		Background	131
	6.1.1	Paper Charts	131
	6.1.2	Types of Electronic Charts	132
	6.1.3	Comparison of Paper and Electronic Charts	132
6.2		General	135
	6.2.1	Transition from Paper to Electronic Charts	135
	6.2.2	Updates to Electronic Charts	137

	6.2.3	Hard Copy Backups of Electronic Charts	138
	6.2.4	Scale Information	139
	6.2.5	Basic Zooming and Panning	140
	6.2.6	Procedures for Use of Electronic Charts in Part 121/135 Operations	142
	6.2.7	Orientation of Electronic Charts	143
	6.2.8	Using Charts with Other Flight Deck Displays of Spatial Information	145
	6.2.9	Access to Individual Charts	146
	6.2.10	Knowledge and Display of Own-Aircraft Position	148
	6.2.11	De-cluttering and Display Configuration	150
	6.2.12	Color Coding of Chart Symbols	151

7 References 152
Appendix A: EFB Industry Snapshot 155
Appendix B: Summary of Equipment Requirements and Recommendations 181

List of Tables

Table 1-1 Styles that indicate prioritized guidance and design-tradeoff information. 3
Table 3-1. Differences between advanced and basic EFB operating systems. 74
Table 3-2. Comparison of advanced and basic electronic document features. 75
Table 4-1. Comparison of paper and electronic checklist features (continued on next page) 94
Table 4-2. Comparison of paper and electronic checklist features (continued from previous page). 95
Table 6-1 A comparison of paper and electronic chart features (continued on next page). 133
Table 6-2 A comparison of paper and electronic chart features (continued from previous page). 134

Executive Summary

Electronic Flight Bags (EFBs) hold great promise for increasing efficiency and safety of flight operations. However, EFBs could also have negative side effects if not implemented appropriately. For example, they could increase workload and head-down time, and distract the flight crew from higher priority tasks. These potential negative impacts of EFBs need to be addressed during both design and evaluation.

In order to understand and assess the full impact of an EFB, designers and evaluators require an understanding of how the device will function and be used by crews, how the device will interact with other flight deck equipment, and how training and operating procedures will be affected. The purpose of this report is to identify and prioritize guidance on these topics so that designers and evaluators can make informed choices. In the report, we highlight established user interface design principles, recommendations, and design tradeoffs for these systems. Sources for more information (e.g., FAA regulations and other industry documents on best design practices) are referenced where appropriate. The material in this document applies generally to all operations (Part 91, Part 121, Part 125, and Part 135), and to all varieties of EFBs, except when noted.

This document, "Version 2," supersedes an earlier report from September 2000, subtitled "Version 1: Basic Functions" (DOT-VNTSC-FAA-00-22), which is cited in the FAA's Advisory Circular (AC) 120-76A. Version 2 is more comprehensive and detailed than the earlier report. Although this document is not regulatory, we do highlight guidance statements that the authors feel are minimum acceptable practices. Regulatory application of information in this document is the responsibility of the FAA or other appropriate regulatory government agencies.

Much of the guidance in this document is general and applies to any EFB system, regardless of the applications that are supported. The general guidance covers integration with other flight deck systems, physical considerations related to placement of the EFB in the flight deck, general software and hardware topics, and general training/procedures topics. Application-specific guidance is provided for electronic documents, electronic checklists, flight performance calculations, and electronic charts. Other applications that have recently become more prominent, such as cabin video surveillance, are not addressed directly in this report, but the general guidance still applies. Human factors guidance currently being drafted for other applications, such as surface moving map and weather applications, can be found in documents by industry standards committees such as RTCA.

Two appendices are included in this document. Appendix A (EFB Industry Snapshot) contains information about the EFB industry as it stands in September of 2003. EFB systems, hardware, and software products are described, including hyperlinks to company and product websites. Appendix A should be helpful to anyone needing a quick overview of the EFB industry, which is growing and changing rapidly. Appendix B (Summary of Equipment Requirements and Recommendations) contains a summary of the high priority guidance that would be useful for equipment evaluations. It is designed for use by manufacturers and regulators when conducting office-setting human factors evaluations of EFBs.

Acronyms

AMJ	Advisory Material Joint
ATIS	Automatic Terminal Information System
AQP	Advance Qualification Program
ARP	Aerospace Recommended Practice
CAP 708	Guidance on the Design, Presentation and Use of Electronic Checklists (See References)
CDU	Control and Display Unit
CFR	Code of Federal Regulations
ECL	Electronic Checklist
EICAS	Engine Indication and Crew Alerting System
EFB	Electronic Flight Bag
EPGWS	Enhanced Ground Proximity Warning System
FAA	Federal Aviation Administration
FMC/FMS	Flight Management Computer/Flight Management System
GUI	Graphical User Interface
HFDS	Human Factors Design Standards for Acquisition of Commercial-off-the-shelf Subsystems, Non-Developmental Items, and Developmental Systems (See References)
IFR	Instrument Flight Rules
JAA	Joint Aviation Authorities
LOE	Line Oriented Evaluation
MEL	Minimum Equipment List
PDA	Personal digital assistant
PDF	Portable Document Format, a page-definition language
PFD	Primary flight display
POH	Pilot Operating Handbook
PAI	Principal Avionics Inspector
POI	Principal Operations Inspector
SAE	Society of Automotive Engineers
VFR	Visual Flight Rules

1 Introduction

Electronic Flight Bags (EFBs) are coming into the flight deck and bringing along with them a wide range of human factors considerations. It is clear from the EFB Industry Snapshot (Appendix A) that there are a large variety of EFB products on the market as of September 2003, and that the industry is flourishing.

In order to understand and assess the full impact of an EFB, designers and evaluators require an understanding of how the device will function and be used by crews, how the device will interact with other flight deck equipment, and how training and operating procedures will be affected. The purpose of this report is to identify and prioritize guidance on these topics so that designers and evaluators can make informed choices. Because this document is lengthy, we also provide a summary version in Appendix B, which can be used by either manufacturers or regulators as a quick-reference guide for evaluating EFB equipment in an office setting. Note that Appendix B excludes guidance on installation issues and training/procedures issues, which is provided in the main text.

This document, "Version 2," supersedes an earlier report from September 2000, subtitled "Version 1: Basic Functions" (DOT-VNTSC-FAA-00-22), which is cited in the FAA's Advisory Circular (AC) on EFBs, AC 120-76A. In this report, we highlight established user interface design principles, recommendations, and design tradeoffs for these systems. Sources for more information (e.g., FAA regulations and other industry documents on best design practices) are referenced where appropriate. The material in this document applies generally to all operations (Part 91, Part 121, Part 125, and Part 135), and to all varieties of EFBs, except as explicitly noted.

Much of the guidance in this document is general and applies to any EFB system, regardless of the applications that are supported. The general guidance covers integration with other flight deck systems, physical considerations related to placement of the EFB in the flight deck, general software and hardware topics, and general training/procedures topics. Application-specific guidance is provided for electronic documents, electronic checklists, flight performance calculations, and electronic charts. Other applications that have recently become more prominent, such as cabin video surveillance, are not addressed directly in this report, but the general guidance still applies. Human factors guidance currently being developed for other applications, such as surface moving map and weather applications, can be found in documents being drafted and published by the various standards committees.

1.1 Background

More than four years ago, industry and the FAA recognized that human factors concerns would play a key role in the design and evaluation of EFBs. While EFBs may look like familiar equipment in an office setting, they are new and sophisticated devices from a flight deck perspective. For example, many of them have graphical user interfaces and can support multiple new functions, such as electronic charts and documents. So when the industry first sought guidance from the FAA on EFB technology, Volpe Center was tasked with identifying and prioritizing EFB human factors considerations.

Working with the FAA and the Air Transport Association Digital Display Working Group, Volpe produced a lengthy document on EFB human factors considerations (Chandra and Mangold, 2000). That report, known as Version 1, is referenced in the FAA Advisory Circular (AC) on EFBs (AC 120-76A), which was issued on March 17, 2003. In addition, topics from the Version 1 report that were considered especially important are brought into the main text of the EFB AC. The EFB AC provides a relatively streamlined approval and certification process for EFBs that meet specific criteria related to installation and functionality.

Like the earlier report, this updated Version 2 document contains information for FAA evaluators, system designers/manufacturers and operators about the many EFB human factors considerations that need to be addressed during the design and evaluation. Industry comments obtained on the Version 1 report were incorporated into this release. As a result, this Version 2 report is more comprehensive and detailed.

Note, however, that this Volpe Center report is not regulatory. The regulatory application of information in this document is the responsibility of the FAA or other appropriate government agencies.

1.2 How to Use this Document

This document is structured for ease of use by a variety of readers from industry and government. Most readers are expected to browse or search the document for specific topics of interest. As a result, each substantive topic is addressed in short self-contained sections called *considerations*, which are one or two pages long. The intent is to capture, in one document, all relevant material to help a wide international audience of readers, to whom different sections may be more or less obvious.

Each consideration begins on a new page and has a title that describes the general subject. The consideration begins with guidance statements, and then has three labeled subsections: Problem Statement, Examples, and Evaluation Questions. The supporting material (i.e., a description of the problem, its potential impact, and examples of solutions) is helpful in illustrating the spirit of the guidance. The Evaluation Questions are open-ended examples of questions that an evaluator (e.g., FAA certification specialist) could consider; they do not provide detailed guidance on performance assessment, but do point out areas for evaluation.

To browse the document, read just the guidance statements at the top of each consideration. The guidance within each consideration is separated into outlined statements. Each outlined statement is preceded by a two-part descriptive label, such as "Equipment Requirement", or "Installation Recommendation." The first part of the label identifies whether the statement concerns *Equipment* (i.e., the specific hardware or software), *Installation* (i.e., how the unit functions in the context of an aircraft), *Training/Procedures* (i.e., how the crew uses the equipment), or some combination of these topics. *Equipment* guidelines are design items that can typically be assessed in a bench test; i.e., these items are testable outside the context of the aircraft. Some equipment guidelines are more focused on software, while others are more focused on hardware. Note that, a particular EFB may have implemented a particular type of function either through hardware or software, or through some combination, so it is not possible to determine in advance whether the guideline is to be addressed at the hardware or software level. The applicability of each equipment guideline to a given product is determined at the time of the review. *Installation* guidelines are those that need to be evaluated within the context of the aircraft flight deck. *Training/Procedures* guidelines are, of course, related to training and procedures rather than the design or installation of the equipment itself. In some cases, the term is used more broadly to reflect involvement of any personnel, not just the flight crew.

The second part of the label for the outlined guidance statement identifies the priority-level or type of information in the statement. The style of outlining varies as shown in Table 1-1 below. Guidance statements labeled *Requirements* are ones that the authors feel are minimum acceptable practices. In many cases, requirements can be traced back to other references, which are provided. Note that we use the term "should" instead of "shall" within text of the Requirements, to remind the reader that this document is not regulatory. *Recommendations* describe highly preferred methods or mechanisms. Compliance with recommendations produces a better system, but increased cost or lack of feasibility may deter some designers or operators from implementing them. *Suggestions* are options that should be considered, but may not be appropriate in all circumstances. In contrast to requirements, recommendations, and suggestions, which are prescriptive to various degrees, *Design Tradeoffs* are descriptive. That is, we describe tradeoffs without specifying a "correct" or "best" solution.

For more information about that topic, read the rest of the page. In the Problem Statement, we describe the problem that the guidance addresses, including the potential impact if the problem is not addressed. In the Examples section, we list examples of the potential problem and possible solutions. The Evaluation Questions section contains open-ended questions that an evaluator (e.g., FAA certification specialist) could consider when determining whether the problem has been addressed adequately. The Evaluation Questions do not provide detailed guidance on performance assessment, but they do point out areas for evaluation.

Introduction

Requirement

> - Requirements are shaded and outlined with a double line. These are statements that the authors feel are minimum acceptable practices. Although we use the term "requirement" within the document, note that this is not a regulatory document and any application of these requirements is the responsibility of the appropriate regulatory agency (such as the FAA in the United States). We use "should" instead of "shall" in the Requirements language to remind the user that the document is not regulatory.

Recommendation

> - Recommendations are outlined with a thick line. These are highly preferred methods or mechanisms.

Suggestion

> - Suggestions are outlined with a thin line. These are options that should be considered, but may not be appropriate in all circumstances.

Design Tradeoff

> Design Tradeoffs are outlined with a dashed line. These are descriptive (not prescriptive) statements that point out design tradeoffs and other related factors without specifying a solution.

Table 1-1 Styles that indicate prioritized guidance and design-tradeoff information.

2 General EFB System

Considerations in this chapter are general and apply to any EFB system, regardless of the function, or functions, performed by that EFB. These considerations are divided into five sections: General, Physical, Training/Procedures, Software, and Hardware.

2.1 General Considerations

2.1.1 Workload

Equipment Recommendation

> - EFB software should be designed to minimize flight crew workload and head-down time. (See 14 CFR Parts 23.1523, 25.1523, 27.1523, 29.1523, and associated AC 25.1523-1, on Minimum Flightcrew, current version. (AC 120-76A, Section 10.c)

Equipment and Training/Procedures Recommendations

> - The workload associated with using the EFB should be assessed to ensure that it is acceptable. (See AC 25.1523)
> - The EFB workload assessment should include a theoretical analysis, such as a comparison between the new design/procedure and an existing design/procedure. This theoretical analysis may be sufficient to assess workload for routine EFB tasks done under normal or low workload conditions.
> - For demanding tasks performed with the EFB under time pressure, the workload assessment should also include an operational component. In the operational test, workload should be measured under realistic conditions to ensure that it is acceptable. (See 2.1.2)
> - During low workload situations, it may be acceptable for the EFB to increase workload slightly for short periods of time if overall performance and/or workload is improved. If this is the case, procedures should be designed to control additional workload created by using an EFB. (AC 120-76A, Section 10 f (3))
> - When using the EFB for complex tasks, the crew may become distracted by the EFB and neglect higher priority tasks. The risk of this type of distraction from an EFB should be minimized. (See 2.1.2 for options on how to minimize this risk.)

Training/Procedures Design Tradeoff(s)

> It is often difficult to measure workload differences between tasks, particularly because workload is often shifted around by new technology rather than eliminated. Designers and evaluators should be aware of workload tradeoffs. They should ask how the workload is redistributed, and not just whether the workload is decreased or increased.

Problem Statement

When crew workload is too high, or too low, performance suffers. Therefore, it is important to understand how a new system such as an EFB will affect workload patterns. Workload may be decreased in some ways and increased in other ways. Increased workload could result from inefficient design of the software or hardware, or even from limitations in the flexibility of using EFBs in relation to paper documents.

It is important to understand what the new workload patterns will be and to judge in advance whether the new workload levels are acceptable. Note that there may be differences in both physical workload (reaching, manipulating, and viewing) and mental workload (the thought required to identify and find the desired information).

It is difficult to assess workload of a single system. It is more typical to compare the workload of using a new system/procedure against the workload of using a system/procedure that has been demonstrated to have acceptable workload in service.

Workload may be assessed through a theoretical analysis, an operational assessment, or both. The theoretical analysis may suffice in some cases, but the operational assessment may be necessary under more challenging conditions. In the operational assessment, workload could be measured with subjective measures (e.g., Cooper-Harper rating scale) and/or objective measures (e.g., flight performance).

When evaluating the workload incurred by use of an EFB, the evaluator should consider factors such as the time it takes to complete the task using the EFB, as compared with doing the task without the EFB. The location and accessibility of the EFB display and controls, the amount of automation, and the usability of the EFB software will all affect the time it takes to complete a task using the EFB. The evaluator should also consider whether users would be more likely to make errors during high workload conditions, whether it is easy to recover from errors, and whether users are likely to become distracted from other flight deck tasks while resolving EFB problems.

Examples

An EFB may provide flight crews with a new capability, such as completing weight and balance calculations. This new responsibility may be in addition to the other tasks that the flight crew is used to performing, so in a sense it is an increase in the flight crew's workload. Procedures should ensure that the workload associated with this type of new task is acceptable. For example, crews could be allowed to update weight and balance computations only while at the gate, rather than during taxi, or they could use these functions only to review or modify calculations while taxiing.

The workload required to manipulate electronic documents may exceed the workload required to manipulate paper documents. Although workload might increase with electronic documents, this negative quality is offset by other factors, such as the improved electronic search capabilities, and the fact that documents are typically referenced in low workload conditions. Overall, the net increase in workload may be judged acceptable.

Instead of supporting new tasks, an EFB may allow flight crews to perform existing tasks more efficiently, such as looking up reference information from a flight manual. In this case, the design of the software search procedure can affect the risk of getting lost in the process of searching for information, or the risk of becoming distracted by a search that results in too many choices. An appropriate design of the search procedure should mitigate these risks.

It may be hard to find a good viewing position for a portable EFB that shows electronic charts. The EFB is less flexible than paper in this sense. The reduced flexibility of positioning an EFB may affect the pilot's task by increasing head-down time, and as a consequence workload.

Evaluation Questions

- How does the workload required for completing a task with the EFB compare with the workload for completing the task with an alternative approved method? If there is an increase in the workload of completing a task with the EFB relative to alternative methods, is this increase acceptable?
- Does the software and/or hardware design minimize any increase in workload?

2.1.2 Using EFBs During High Workload Phases of Flight

Equipment, Installation, and Training/Procedures Recommendations

> - If the EFB is used in high workload phases of flight, such as take-off and landing, its use should be evaluated through simulations and actual flight testing under those conditions. (AC 120-76A, Section 10.c)
> - Under high workload conditions, the procedures for using the EFB, its hardware and/or software design, and even the installation of the device should be designed to minimize workload. (See AC 25.1523)
> - Unsecured EFBs are not recommended for use in high workload conditions for two reasons. First, there is extra workload to ensure that the unsecured device does not move about unexpectedly, potentially jamming controls and causing damage. Second, there is additional workload in positioning the unsecured device for proper viewing and use. (See 2.2.3)

Training/Procedures Recommendation

> - Complex, multi-step data entry tasks should be avoided during takeoff, landing, and other high workload phases of flight. (AC 120-76A, Section 10.c)

Equipment, Installation, and Training/Procedures Design Tradeoff(s)

> EFBs could distract crews from higher priority tasks; this is especially undesirable during high-workload phases of flight. Procedures, policies, and perhaps even built-in limits on the use of the EFB may be useful in addressing this problem, particularly for EFBs that have knowledge of aircraft system status (e.g., discourage access to non-flight applications in flight). (See 2.1.1)

Problem Statement

Using an EFB requires effort that may be different from that of using paper. There may be effort involved in locating and orienting the display for use and there is effort in looking at the display, processing the information, and making any necessary entries. Data entry can produce particularly long head-down times and high workload. Visual scanning of the EFB (without data entry) does not require as much effort, but may still be an additional task for the pilot, depending on the function.

The additional workload required to use an EFB may distract the pilot from higher priority time-critical tasks, which is particularly detrimental during high workload phases of flight. Ultimately, adding workload to high workload phases of flight could potentially affect the certification basis under 14 CFR Part 25.1523, requiring re-evaluation of compliance with regulations.

Examples

During high workload situations, such as takeoff and landing, entering data on the EFB may distract the crew from essential functions, such as visual scanning for air traffic out the window, or scanning of aircraft instruments. Data entry tasks should be avoided during these phases of flight. If data entry is required, it should be limited to a single key press. For example, to indicate that the climb-out checklist has been completed, the pilot may enter a yes/no response to an EFB inquiry.

If, however, the EFB is used as a display of real-time information useful during landing (e.g., if the EFB displays nearby traffic during landing), and only requires occasional scanning that the pilot can incorporate into his/her task schedule, the additional workload may be acceptable. An operational evaluation may be necessary to ensure this conclusion.

An EFB that has more built-in automation may also be more acceptable for use under high workload conditions. For example, if some items in an emergency checklist are completed through aircraft sensors, the pilot's workload may not be impacted negatively by using the EFB as compared with the paper checklist.

Some EFBs that have knowledge of aircraft system status may have built-in limits, such as the inability to exercise certain functions below 10,000 ft altitude.

General EFB System

Evaluation Questions
- Are complex, multi-step data entry tasks avoided during takeoff, landing, and other high workload phases of flight? Does the use of the EFB impose additional workload during a high workload phase of flight? If so, did the operational evaluation confirm that the overall workload is acceptable?
- If the EFB will be used under high workload conditions, has its use been evaluations in an operational assessment?
- If the EFB is designed for use during high workload phases of flight, is it secured within the aircraft?
- What training and/or procedures are in place to mitigate any additional workload of using the EFB under high workload conditions?
- Are there procedures, policies, or built-in limits on use of the EFB to ensure that pilot do not become distracted during high workload phases of flight?

2.1.3 Compatibility and Consistency with Flight Deck Systems and Other Flight Information

Installation and Equipment Recommendations

> NOTE: For guidance on internal EFB application consistency, see 2.4.2.
>
> - The user interface of the EFB should be compatible with the crew interface and flight deck design philosophy of the particular aircraft in which it will be used in order to minimize flight crew error and maximize the identification and resolution of errors. (AC 120-76A, Section 10.e, (1))
>
>   NOTE: In order to be *compatible*, the EFB user interface should not be in direct conflict with other systems (e.g., it should not present information that conflicts with information from other flight deck systems).
>
> - Data entry methods, color coding philosophies, and symbology should be as consistent as possible between the EFB and other flight deck systems and paper documentation (such as dispatch papers, weather reports, etc.). (AC 120-76A, Section 10.b (1) and Section 10.e (1))
>
>   NOTE: In order to be *consistent*, the EFB user interface should be identical to the other systems (e.g., it should be able to present information in the same way as it is presented on other flight deck systems).

Equipment Design Tradeoff(s)

> In order to be both compatible and consistent with multiple flight deck systems, which may have inconsistencies among themselves, it may be necessary for the EFB to support multiple conventions, e.g., for data entry.
>
> In general, the pilot should only have to enter data into flight deck systems once. However, if the EFB does not have access to the other flight deck systems, it may be necessary to make duplicate entries into the EFB (e.g., flight plan data). Designers should consider how to make the EFB data entry as simple and error free as possible given that the pilot may be duplicating other entries.

Problem Statement

Today's complex, integrated flight decks are designed with an overall flight deck design philosophy in mind. The design philosophy guides system development in important and sometimes subtle ways to enhance overall simplicity and safety. While pilots may not be able to state what the flight deck design is verbally, they do develop implicit expectations of how the aircraft will operate based on their experience with that flight deck.

When a new system such as the EFB is introduced into an existing flight deck, it is important that the pilot's expectations of how the aircraft operates are not violated by the EFB. If the EFB is incompatible with the overall flight deck design, pilots will have trouble learning to use it, and they will be more likely to make errors.

This is also true of the information generated by the operator and its service providers. Airlines maintain documentation and information services that rely on particular formats for information, and information provided by the EFB should be consistent with them.

Examples

One design philosophy, implemented by the Boeing Company, is that the flight deck should normally be "quiet and dark." By this, designers mean that the flight deck should (a) be as dark as possible during night-time operations so that the pilots' eyes become dark adapted, easing the transition between outside and inside viewing, and (b) in general, there should be few or no lit status indicators or audible alerts during normal conditions. To be consistent with the quiet/dark philosophy, EFBs should also be designed with dark displays (i.e., black backgrounds with lighted text), and they should have few or no visual status indicators or associated audible alerts in normal operations.

Examples of other sources of flight information that the EFB should be consistent with include dispatch information, maintenance documentation, and weather forecasts. These documents use units of measurement, abbreviations, and symbols that should be used consistently on the EFB.

Wind vector input in the FMC may require a particular format. However, the EFB could accept that format, and other formats that also clearly represent wind data. In this case, the system would be compatible with the flight deck system, while allowing even more flexibility.

Evaluation Questions

- Is the user interface of the EFB compatible with the philosophy of the flight deck?
- Does the EFB minimize the potential for crew error by using terms, color codes and symbols consistent with flight deck systems and other sources of flight information?

2.1.4 Use of EFB with Other Flight Deck Systems

Training/Procedures Requirements

> - Procedures should be designed to ensure that the crew knows what flight deck automation system (e.g., EICAS, FMS, or EFB) to use for a given purpose, especially when information/data is provided by both aircraft and EFB systems. (AC 120-76A, Section 10.f (1))
> - Procedures should also be designed to define actions to be taken when information provided by an EFB does not agree with that from other flight deck sources, or when one EFB disagrees with another. (AC 120-76A, Section 10 f (1))
> - If an EFB generates information that existing flight deck automation also generates, procedures should be developed to identify which information source will be primary, which source will be used as a backup, and under what conditions to use the backup source. (AC 120-76A, Section 10.f (1))

Problem Statement

Whether or not there is any communication between aircraft systems and the EFB, from the perspective of a crew member, the EFB is just another tool for him/her to use. If there are inconsistencies or redundancies in the information provided by the different automation systems ("tools") or information sources, there will be confusion and increased potential for errors.

Therefore, regardless of whether there is a data connection between the flight deck systems and the EFB, information consistency/redundancy should be considered when integrating an EFB into an aircraft with other advanced systems, such as a Flight Management System (FMS) or Engine Indication and Crew Alerting System (EICAS).

Examples

EFBs may support electronic checklists. On some aircraft, both electronic checklists and paper checklists are provided. Procedures should be established to ensure that crews know which of these checklists should be used, particularly in an emergency.

The EFB and FMC could both compute performance data separately. Crews need to know which performance data are to be used for the flight, and they should ensure, if appropriate, that these are the data in use by automation on the aircraft (e.g., the FMC and/or auto-pilot). Performance data disagreement between an EFB and an FMS may cause the pilot to set speed bugs to incorrect values for autothrottle-controlled takeoffs and may result in sub-optimal cruise speeds or surprising FMS behaviors, such as unexpected predicted speeds, estimated times en route, and top of descent points.

Evaluation Questions

- Are there procedures to ensure that the crew knows what flight deck system information is to be used if there is any redundancy with the information from the EFB?
- What procedures does the crew follow if there is a disagreement between the EFB and other flight deck systems, or between multiple EFBs? What are the consequences of using backup information?
- What are the procedures for establishing which source of information is primary?
- Under what conditions will the backup source of information be used?

General EFB System 13

2.1.5 Legibility—Lighting Issues

Installation Requirement

> - The EFB display should be legible to the typical user under the variety of lighting conditions expected in a flight deck, including use in direct sunlight. (See AC 25.11)

Installation and Equipment Requirement

> - Users should be able to adjust the screen brightness of an EFB independently of the brightness of other displays in the flight deck.

Equipment Requirement

> - If automatic brightness adjustment is incorporated, it should operate independently for each EFB in the flight deck.

Equipment Recommendations

> - Screen brightness should be adjustable in fine increments or in a continuous rather than discrete manner.
> - Buttons and labels should be adequately illuminated for night use.

Equipment Design Tradeoff(s)

> There may be special considerations for EFBs to be used with other devices, such as head-up displays or night vision goggles, where additional lighting issues need to be considered.
>
> Standards for the range of brightness that should be supported by an EFB have not been determined. In general, a broad range of brightness is highly desirable in order to maximize display usability in both nighttime and bright daylight operations. Note, there is guidance on brightness for CRT displays in AC 25.11.

Problem Statement

Ambient lighting conditions may vary from very dark during a night flight to direct sunlight on the display. The EFB display should be usable under all of these lighting conditions. This will require at least some ability to adjust the screen brightness. In addition to screen brightness, the size of the text, font style, contrast, glare/reflections, display quality, colors, viewing distance, and off-angle viewing will all affect text legibility. If the EFB is portable, legibility should be checked both inside and outside the flight deck.

Screens or text that are not legible will cause pilot distraction at the least (as the pilot attempts to position the display for better legibility) and potentially more harmful consequences if important information is misread, or not read at all.

Examples

Different font styles may be used at the same time on an EFB. The different fonts may indicate something about that information. For example, one font style may denote the active checklist item (e.g., white, large font) and another font style could represent a completed item (e.g., green, small font). Because the font styles can encode important information, all of the font styles on the screen should be legible and easily discriminated from one another without any screen adjustments.

Tests for readability under different lighting conditions are listed in AC 25.11.

Evaluation Questions
- Can the EFB screen be read under a variety of typical flight-deck lighting conditions? Can the EFB screen be read outside the flight deck?
- Can the user adjust the screen brightness and contrast (if applicable)?
- Are buttons and labels adequately illuminated for night use?

General EFB System 15

2.1.6   System Shutdown

Installation, Equipment, and Training/Procedures Requirements

> - Shutdown procedures for EFBs should be designed such that (AC 120-76A, Section 10 f (6)):
>
>   (a) flight crews can incorporate it into their normal aircraft shutdown procedures without undue difficulty, and
>
>   (b) the EFB operating system remains stable after multiple start-ups and shutdowns in the aircraft

Problem Statement

Crews routinely shut down airplane systems by turning off the power supply; they do not expect to wait for systems to complete their shutdown routines before cutting the power. However, some EFB operating systems need to perform shutdown routines before the power is cut in order to work properly the next time they are started. If the power is cut abruptly, the EFB operating system may be corrupted and fail to function.

Examples

Standard desktop operating systems such as Windows perform shutdown procedures that take several seconds, if not minutes to complete. For continued stable use in the flight deck, the shutdown procedures for a Windows-based EFB should be initiated well in advance of shutting down power to the aircraft.

EFBs that are aware of aircraft status could detect when they should begin shutdown routines. For example, when the parking brake is set, and the engines are all out and a passenger door is opened, the EFB could begin shutdown routines without crew input.

Evaluation Questions

- What are the procedures for shutting down the EFB?
- Are the shutdown procedures designed for long-term stability of the EFB and ease of crew operation?
- What happens if the crew cuts power to the EFB instead of shutting it down properly? Does the EFB function correctly when rebooted?

2.1.7 Failure Modes

Equipment Recommendation

> - The EFB system should be capable of alerting the flight crew of probable EFB application/system failures. (AC 120-76A, Section 10.e (2))

Equipment and Training/Procedures Recommendation

> - The effects of undetected errors in all EFB applications should be evaluated for each application. The assessment should address the adequacy of the human/machine interface, accessibility of controls, ability to view controls, annunciations, displays and printers, and the effect on flight crew workload and head-down time. The assessment should also consider the effects of flight crew (procedural) errors determined by comments from the professional pilot community. (AC 120-76A, Section 10.e (2))

Training/Procedures Recommendation

> - The procedures to follow when one or more unit fails should be clear to the flight crew.

Equipment Suggestion

> - EFB failures should be graceful. For example, if one application fails on the EFB, other applications should still function normally. Also, warm (software) reboots should generally recover the system after a failure. Cold reboots should be atypical.
>
> - Depending on the nature of the EFB application/system failure, pilot notification and/or acknowledgment may be appropriate (see also 2.4.9). A state transition diagram may be useful for identifying how each failure, or partial-failure, condition will be indicated to the crew and whether crew acknowledgement is appropriate.

Problem Statement

There is a need to identify and classify failure modes and effects for each EFB application. The purpose of this analysis is to evaluate and plan for the effects of undetected errors in all EFB applications. To perform this analysis, errors in the EFB applications should be assumed, and the consequences of these errors and procedures for recovery should be studied. Failures should be designed to be graceful, in the sense that they can be recovered from easily, with minimum disruption to flight crew tasks and workload. If failures are not easily recognized, if failures are difficult to recover from, or if procedures for handling failures have not been developed in advance, crew workload and performance may suffer significantly at the time of an EFB failure.

Examples

Failure modes for applications that present static data (e.g., electronic documents) are generally limited to loss of access to the data. However, functions that are interactive and dynamic (e.g., presentation of ownship location on an electronic chart) are subject to other failures, such as complete loss of a required data source (e.g., GPS position) or loss of integrity of the required data source.

A performance calculation application could potentially produce incorrect results that might be difficult for the crew to recognize. This anticipated failure mode puts a constraint on the data integrity checking of the performance database. If only validated data are loaded, such errors cannot occur.

In an electronic chart application that is aware of ownship position and orientation (e.g., track or heading), there could be independent or joint partial failures, such as loss of position, or track, or heading data. Each of these failures and their potential combinations may call for different types of crew notification and input. A state transition diagram could be useful for clarifying these different situations.

Evaluation Questions
- Are failure modes obvious to the crew? Is the nature of the failure clear?
- Are failures handled gracefully, with minimum impact to crew tasks and workload?
- Are there procedures in place for the crew in case a failure occurs?
- If the EFB hangs and fails to respond to crew input, or displays error or fault messages, are the means of recovery easy to remember and apply? Does the crew have to remember any arbitrary procedures or refer to paper documentation in order to restart the EFB?

2.2 Physical Considerations

2.2.1 Design and Placement of Structural Cradle

Installation and Equipment Requirements

> NOTE: A structural cradle is a piece of hardware that is (a) physically attached to the aircraft and (b) designed to hold the EFB while it is in use.
>
> - If a cradle is added to the aircraft, it should meet 14 CFR Parts 25.561, 25.562, 27.561, 27.562, 29.561, or 29.562 (as appropriate) regarding safety during dynamic or static emergency landing conditions.
>
> - When the EFB is in use and is intended to be viewed or controlled, it should be within 90 degrees on either side of each pilot's line of sight. (AC 120-76A, Section 10.b (3))
>
> Note: A 90° viewing angle may be unacceptable for certain EFB applications if aspects of the display quality are degraded at large viewing angles (e.g., the colors wash out or the color contrast is insufficient at the installation viewing angle). Each EFB should be evaluated with regard to viewing angle requirements.
>
> - The structural cradle should not be mounted such that it obstructs visual or physical access to controls and/or displays, flight crew ingress or egress, or external vision. (See 14 CFR Parts 25.1321 and 29.1321) (AC 120-76A, Section 10.b (2))

Installation Recommendations

> - If an EFB is being used to display high priority flight information such as for navigation, terrain and obstacle warnings that require immediate action, takeoff and landing V speeds, or for functions other than situation awareness, then such information should be in the pilot's primary field of view, well inside the 90° viewing angle requirement. (AC 120-76A, Section 10.b (3))
>
> - The cradle should allow easy access to all EFB controls and a clear view of the EFB display. (AC 120-76A, Section 10.b (2))
>
> - The structure and associated mechanism should not impede the flight crew in the performance of any task (normal, abnormal, or emergency) associated with operating any aircraft system. (AC 120-76A, Section 10.b (2a))
>
> - The cradle should be mounted such that the pilot does not have to turn significantly to use the device, particularly during high workload phases of flight. The cradle should also be easily accessible by a wide population of users. It should not be placed in a space where it is awkward to reach or manipulate the EFB.
>
> - The cradle should allow for some adjustment of the EFB orientation to allow pilots to (a) customize the viewing angle to some extent and (b) alter the pattern of ambient light that falls on the display to reduce significant glare and reflections, which could impact its readability.
>
> - Cradle structures should be able to lock in position easily. Selection of locking positions should be fine enough to accommodate a range of crewmember preferences. In addition, the range of available selections should accommodate the expected range of users' physical abilities (i.e., anthropometrical constraints). (AC 120-76A, Section 10.b (2b))
>
> - The cradle should allow the user to lock the EFB in a position that does not interfere with flight crew operations when not in use. The cradle should be mounted so that the EFB is easily accessible when stowed. (AC 120-76A, Section 10.b (2c))

General EFB System   19

Installation Recommendations (continued)

> - The installation of the EFB should minimize any additional workload incurred by the need to locate and orient the display, particularly if used during high workload phases of flight. (See 2.1.1 and 2.1.2)

Training/Procedures Requirements(s)

> - Crews should know how and when to adjust the cradle position, their own seating position, or the placement of the EFB unit such that they have easy access to, and a clear view of, the EFB.
> - Procedures should ensure that pilots have good access to all flight controls and displays, even those that are partially obstructed by the EFB and its cradle when necessary.

Problem Statement

Devices that are added into older flight decks could obstruct access to or use of other equipment. A structural cradle can help assure that the device is positioned appropriately for use.

The pilot should not be required to use an EFB that is located behind him/her because that position is more likely to distract him/her from important tasks, such as scanning out the window for traffic.

Examples

The structural cradle could be on the yoke of the aircraft. In this case, the pilot needs to adjust his/her seat position in order to have a clear view of the EFB and other flight deck displays.

Pilots who are unfamiliar with an approach into a given airport might prefer to have the EFB electronic chart on the control column so he or she can learn the procedure while following flight director guidance. If the EFB cannot be mounted in front of the pilot as a paper chart can, it may make following a procedure more difficult. The pilot would have to look to the side instead of to the front. This may result in more head down time and higher workload, including higher mental workload to associate the chart depiction with the pilot's current position in the procedure.

The structural cradle could be positioned to the side of the pilot, just outside the existing flight deck panel. Again, the pilot may have to adjust his/her seat to be at the reference eye position.

Evaluation Questions

- Does the structural cradle obstruct visual or physical access to flight controls and/or displays? Which controls/displays are affected, and how important are they during the different phases of flight in which the EFB will be used?
- Do crews know how to adjust and lock the EFB or their own orientation for optimal viewing or for stowage? Is there adequate room to manipulate the device controls and view its display?
- Is the EFB installed for easy access if used during high workload flight phases?

2.2.2 Stowage Area

Installation Requirements

- A stowage area with a securing mechanism for the EFB is required for storage of portable EFBs when they are not in use. (AC 120-76A, Section 11.f)

  NOTE: If the EFB is designed to be held in a structural cradle, the cradle may satisfy the requirement for a stowage area.

  NOTE: For EFBs that are used on light aircraft operating under Part 91, other than subpart F, this requirement may be downgraded to a recommendation.

- The stowage area should be in compliance with 14 CFR Parts 25.787 and 25.789.
- A stowage area is also required for any EFB component that is not fixed in place, e.g., an external keyboard.

Installation Recommendations

- The stowage area and securing mechanism should not be mounted such that they obstruct visual or physical access to controls and/or displays, flight crew ingress or egress, or external vision.
- The EFB should be stowed such that is easily accessible by a wide population of users. In other words, the crew workload of locating, retrieving, and orienting the display should be acceptable. The device should not be in a space that is awkward to reach, or of such a weight and size that it is difficult to manipulate in the given space. (AC 120-76A, Section 10.c)
- The stowage area should be designed such that neither the device (or component) nor the stowage area is easily damaged in normal usage.

Problem Statement

Flight deck real-estate (not just display space) is extremely limited. Every device routinely used in the flight deck should have a designated place, when both in and out of use. Stowage areas should be accessible to crew members without interfering with normal or emergency flight tasks.

The stowage area provides a means of securing EFBs that are unsecured during normal use, including both self-contained devices and those that are tethered to the aircraft.

Examples

EFB units may move unexpectedly during significant accelerations. For example, a unit left on an unused seat may fall off the seat during turbulence. The next time the pilot attempts to use the device, finding the unit will cause pilot distraction at the least.

During takeoff and landing, the EFB may need to be stowed in order to prevent injuries to the crew in case of sudden aircraft accelerations, similar to the requirement for stowing tray tables for passengers.

Evaluation Questions

- Is there a stowage area for the EFB? When the EFB is not stowed, is the securing mechanism in the stowage area unobtrusive?
- When the device is stowed, does the combination of it and the securing mechanism intrude into any other flight deck spaces, causing either visual or physical obstruction of important flight controls/displays?
- Does movement of the EFB to and from a stowage area require substantial effort, or substantially limit access to flight displays and controls? Is the securing mechanism simple to operate for a wide population of users? Is the device or the stowage area easily damaged under normal usage?

General EFB System 21

2.2.3 Use of Unsecured EFBs

Equipment and Training/Procedures Recommendations

> - EFB systems that are not secured in a structural cradle while in use should be designed and used in a manner that prevents the device from jamming flight controls, damaging flight deck equipment, or injuring flight crewmembers should the device move about as a result of turbulence, maneuvering, or other action.

Training/Procedures Recommendations

> - While in flight, crews should routinely store unsecured and tethered EFBs that are not actively in use.
> - While in use, EFBs that are not secured in a structural cradle should not be routinely placed such that they obstruct access to flight controls/displays.

Problem Statement

Unsecured EFBs are those that are either (a) portable and unattached to an aircraft mounting device (i.e., Class 1 hardware as defined in AC120-76A) or (b) tethered units, which are connected to the aircraft via a cable, but whose display unit is not attached to an aircraft mounting device (i.e., structural cradle). Both portable and tethered devices are unsecured because the whole display can be physically manipulated while in use.

Being able to manipulate the position of the EFB display has the advantage that the display can be positioned for maximal utility for any given task. For example, the display could be positioned such that bright sunlight does not fall on it, or the display could be handed over to another pilot, facilitating crew briefing.

However, unsecured EFBs that do not have designated storage and use locations can be a hazard for a number of reasons:

1) They may obstruct access to other displays/controls.

2) In the case of strong accelerations, such as those in takeoff, landing, and turbulence, an unsecured EFB could fall and jam rudder pedals or limit aft yoke travel. Unsecured units could also cause physical injury to the crew under these conditions.

3) There could be confusion when crews attempt to locate, orient, and use portable, unsecured EFBs. This problem will be especially pronounced if the EFB is physically large enough, relative to the size of the flight deck, so as to be difficult to move about quickly and easily.

Examples

An EFB might be placed on the pilot's lap during takeoff. If so, the pilot should insure that he/she has full control authority, i.e., that the yoke can be pulled back completely and the control wheel rotated to full travel without the EFB getting in the way.

Evaluation Questions

- Does the pilot have adequate access to flight controls and displays when the unsecured EFB is in use?
- Where are unsecured EFBs placed when in use, and when out of use?

2.2.4 Kneeboard EFBs

Equipment Requirement

> NOTE: Kneeboard EFBs are those that are secured to the pilot's leg rather than cradle-mounted or hand-held.
> - The kneeboard EFB should be easily removable for emergency egress.

Installation and Training/Procedures Requirements

> - Kneeboard EFBs should allow the pilot to exercise full control authority.
> - Pilots should follow a procedure to check that the EFB allows them full control authority prior to each flight.

Installation Recommendation

> - The securing mechanism holding the kneeboard EFB should be designed for both pilot comfort and convenience in attaching/removing.

Training/Procedures Recommendation

> - Crew should know what they will do with the EFB in an emergency landing. (Keeping the unit attached to their leg may not be the safest option.)

Equipment and Installation Design Tradeoff(s)

> Head-down time associated with use of a kneeboard EFB should be carefully evaluated. Spatial disorientation may be more likely with such a design, especially if the pilot has to look down during while accelerating (e.g., while turning).
>
> Kneeboard EFBs may be in physical contact with the pilot throughout the flight. The heat generated by these devices should be evaluated to ensure that it is acceptable. Also, the weight of the unit on the pilot's leg may be fatiguing after some time.

Problem Statement

Pilots often strap clipboards with paper charts and flight information to their knees. There may be some EFB designs based on this familiar configuration. When pilots use clipboards for paper information, they are responsible for ensuring that the clipboard is properly positioned and does not limit their control authority prior to every flight. The same should be true for any electronic device strapped to the pilot's knee.

Electronic devices differ from clipboards in that they typically weigh more, and they generate heat. It is important that the effects of these factors are mitigated for the comfort of the crew. Other negative aspects of using a kneeboard EFB include an increased risk of spatial disorientation, and the potential for pilot injury from the EFB in an emergency landing.

Examples

Use of kneeboards is particularly common in single-pilot operations (e.g., small general aviation, or military operations). The kneeboard is typically the size of an approach chart and weighs less than 1 lb (approximately 450 g). It holds a few approach charts and notes on weather conditions, air traffic clearances, etc. An electronic version of such a device may need to be a smaller physical size in order to meet the weight constraint.

Kneeboards are often used to take hand-written notes in general aviation operations. An electronic device that replicates this functionality may need to support hand-writing recognition, and have a suitable storage place for a stylus.

General EFB System   23

If the aircraft has an ejection seat, the design of the EFB will have to work properly in the context of an ejection maneuver. The EFB may need to have an automatic quick-release feature. If there are any cables between the EFB and the aircraft, the possibility of entanglements will have to be explored and mitigated.

Evaluation Questions
- Can the kneeboard EFB be positioned such that the pilot has full control authority?
- Is the kneeboard EFB comfortable for the pilot to wear under normal conditions?
- What special procedures are in place for removal of the EFB during emergency landing or egress?

2.3 Training/Procedures Considerations

2.3.1 Part 121, Part 125, and Part 135 Operations EFB Policy

Training/Procedures Requirements

> - Part 121 and Part 135 operators should have a policy that defines how the crew is expected to use the EFB. The policy should cover the specific use of each EFB function during ground operations and under all flight conditions including normal, abnormal, and emergency use.
> - The policy should be documented and provided in written form to flight crews. The operator will need to determine where this policy should be published; it could be included within an existing crew publication, or issued as a separate policy notice.

Training/Procedures Recommendation

> - The EFB policy should address crew resource management (CRM) and crew coordination issues.
> - The EFB policy should be provided to maintenance staff, dispatchers, and other employees whose responsibilities overlap with the functionality supported by the EFB.

Problem Statement

An EFB policy is a general explanation of how the EFB is expected to be used during flight operations and other activities. The purpose of a policy is to provide a framework within which procedures for using the EFB can be designed. Using a policy as the basis for procedure development will ensure that the resulting procedures are internally coherent and consistent with related procedures. Comparing a procedure to the underlying policy can aid in identifying discrepancies and conflicts with the policy. Pilots are more likely to conform to procedures developed from an explicit policy. A written description of the policy should be provided to all appropriate personnel, not just flight crews.

The EFB policy should cover issues such as crew coordination and CRM in order to address the distraction potential of EFB and to provide strategies for managing the use of various functions to prevent distraction. Strategies for crew coordination may vary depending on how many crewmembers are in the flight deck.

Examples

An EFB policy could be similar to an operator's automation philosophy. It could describe the value the carrier expects to gain from the EFB and the role the EFB is expected to play in line operations (flight phases in which the EFB is to be used, etc.). It could also address expected changes in the duties of maintenance, dispatch, and other staff affected by the adoption of an EFB. To be complete, the policy should address each type of functionality that is supported by the EFB. An effective policy reflects the unique operational needs of the carrier.

To address crew coordination issues, the policy should discuss who (pilot-flying or pilot-not-flying) should use the device, and under what conditions. It should also address monitoring and confirmation duties of the crew member who is not actively using the EFB. If two EFB units are on-board, the policy should also address any cross-checking that is required. If the EFB functions duplicate or overlap with other functions or information sources on the flight deck, the policy could describe the operator's philosophy for deciding which information source is primary, and which are secondary.

Evaluation Questions
- Does the air carrier have an explicit policy that addresses the use of the EFB in line operations?
- Is the policy clear and easy to understand and follow? Is it distributed to air carrier personnel?
- Does the policy adequately address each specific EFB application?

General EFB System	25

2.3.2 EFB Documentation for Part 121, Part 125, and Part 135 Operations

Training/Procedures Requirement

> - Existing Part 121, Part 125, and Part 135 operator documentation and policies that could be affected by the introduction of the EFB into line operations should be reviewed and modified as necessary.

Training/Procedures Recommendation

> - The EFB manufacturer should provide explanatory materials on how to use the device to Part 121/125/135 operators. These materials should provide information that is incorporated into the operator's EFB training programs.
> - Adequate documentation should be provided to all EFB users providing guidance on how to use the EFB.

Problem Statement

The successful introduction of new equipment can be aided by the provision of adequate documentation that can be used for training purposes and as a resource for issues that may arise in the future. The EFB manufacturer should provide a starting point for developing air carrier EFB training programs (e.g., in the form of a basic user's manual). EFB information should be incorporated into existing documentation and could be provided additionally as a standalone manual.

Examples

Existing documentation may need to be modified to address EFB use within the larger context of flight operations (e.g., the minimum equipment list). A separate EFB handbook may also be appropriate. In either case, this documentation could include the air carrier's policy on EFB use (see 2.3.1) and an overview of the functionality supported by the EFB. The documentation could also include dispatch relief and procedures, including information about which EFB functions may be inoperable, and if inoperable, what alternative sources and procedures are used.

The logic of the user interface could be described, together with the procedures for using the EFB under normal and non-normal conditions. Standard formatting and color conventions of importance to the end user should be described. Indications of a malfunctioning EFB and procedures for coping with a malfunctioning EFB are also important. A high-level description of the procedures that will be used to upgrade EFB software and content may also be an appropriate topic. Finally, the document should list sources of additional information and help in using the EFB.

Evaluation Questions
- Are other policies affected by the introduction of the EFB modified appropriately?
- Is the documentation provided with the EFB sufficient?
- Did the air carrier incorporate EFB information into its current documentation?

2.3.3 EFB Documentation for Part 91 Operators

Training/Procedures Recommendation

> - Part 91 operators/users should receive explanatory materials on how to use the device from the EFB manufacturer. If the Part 91 operator does not have a company-designed training program, the materials from the manufacturer should be specifically designed for training, and not be just a system specification.

Training/Procedures Suggestion

> - Hands-on training with a qualified instructor may be preferred for some aspects of using the EFB.

Problem Statement

Even "easy to use," well designed EFBs may be mysterious to new users at first, or they may have features that are mysterious even to experienced users. These difficulties may produce inefficient or incorrect use of the device, potentially affecting safety of flight.

Training may not be mandated for Part 91 operators, but a training guide should be provided. Hands-on training with a qualified instructor would be best.

Examples

Applications where interactive data entry is required, and where the resulting computations bear a direct effect on the safety of the flight, may need more formal training than applications that do not affect the safety of flight directly. Also, training is more critical for EFBs that are used as primary sources of information.

Users who are transitioning to new aircraft or to an upgraded model of EFB may need only a differences-training program.

Evaluation Questions

- What materials and/or instruction are provided by the manufacturer on using the EFB?
- What materials and/or instruction are provided by the operator on using the EFB?
- Are the materials clear and easy to understand? Are instructions easy to follow?

2.3.4 Initial EFB Training for Part 121 and Part 135 Operators

Training/Procedures Requirement

> - For Part 121 and Part 135 operators, training programs should ensure that users are competent in operating the EFB, in terms of both their knowledge and skills. (See AC 120-53 for guidance on developing training programs.)

Training/Procedures Recommendation

> - The curriculum (consisting of ground training, desktop simulation, and, if needed, a flight training or aircraft simulator segment) should cover the EFB equipment, operating practices, procedures, and conditions/limitations for use. Discussion of the EFB equipment should include a description of the EFB, its capabilities and applications, a description of EFB controls, displays, symbology (including graphical icons, use of color and standard color coding), and failure modes. Discussion of EFB operating practices should be documented in written form.

Training/Procedures Suggestion

> - Some home study may be an acceptable substitute for classroom study, provided all pilots demonstrate proficiency with the system prior to flight.

Problem Statement

The introduction of new equipment places additional demands on already full training programs. To produce training that is both efficient and effective, minimum knowledge and skill requirements should be defined, and appropriate instruction techniques should be used.

Training will vary based on the variety and criticality of applications on the EFB. Training programs may have to be updated each time the EFB software and/or hardware is updated, unless the change is so small that an internal information bulletin (or equivalent) would be an acceptable substitute. The airline and its principal operations inspector (POI) need to agree on an acceptable training program.

Training may involve classroom or home study. In order to be more efficient, different training programs may be designed for completely new users (e.g., a new airline hire), users who are transitioning to new aircraft, and those who are undergoing annual continuing qualification training.

Examples

Training for initial users may consist of both a separate course on the EFB alone, and integrated training using the EFB in concert with other aircraft systems. For users who are already experienced with the basic device, EFB training could be integrated with more general qualification training. Another option is to incorporate EFB training into the curriculum for a type conversion. For example, if the pilot is transitioning into an aircraft with an installed EFB, the EFB will be just one of many systems they will learn about.

Training needs for EFBs may be more critical for EFBs that are used in high workload phases of flight, or those that are used as the sole means of completing flight-critical functions. For example, if use of the EFB is required during an emergency procedure, Level D training may be required.

Some topics that might be covered in a training program include:

- Company EFB policy
- Retrieving and storing the EFB unit
- Procedures for operating the individual EFB applications, and use of multiple applications at the same time
- Interactions (if any) with other aircraft systems, including all data transactions
- Typical problems that users encounter and strategies for avoiding, detecting, and correcting typical errors and faults

- Procedures for updating EFB data and/or software, and how to check whether the data are approved for use in flight
- MEL status, including paper or other backup
- Crew coordination issues

Evaluation Questions
- Is EFB training designed to meet defined knowledge and skill requirements?
- Is EFB training customized for the audience (e.g., first-time user versus experienced user)?

General EFB System   29

2.3.5   Evaluating EFB Proficiency for Part 121 and Part 135 Operators

Training/Procedures Requirements

> - In order to evaluate EFB proficiency during initial training for Part 121 and Part 135 operators, both the knowledge and skill requirements should be checked.
> - Recurrent training for this group should include appropriate use of the EFB within the context of normal and non-normal procedures.

Problem Statement

Appropriate evaluation of proficiency with EFBs is key to ensuring that pilots achieve at least minimum proficiency during initial EFB training and maintain that proficiency during line operations, as evaluated through both line checks and recurrent or continuing qualification training. The evaluation of proficiency with EFBs should be consistent with the carrier's EFB policy and standard operating procedures.

AQP training, as well as some Part 121 programs, uses a series of evaluation gates that support regular testing and, therefore, early detection of weaknesses in a pilot's knowledge or skills. The number and types of evaluation gates will depend on whether the EFB training is conducted apart from qualification or continuing qualification training or whether it is integrated into existing programs. In either case, evaluating proficiency on the EFB as a part of initial EFB training involves both a knowledge and a skill component.

Examples

Evaluating proficiency with the EFB as a part of recurrent EFB training can take place within the context of appropriate evaluation gates. First-look and maneuver validation are appropriate for evaluating the use of the EFB while executing procedures. The use of the EFB as a decision tool and workload management aid is better addressed during line oriented evaluation (LOE).

Evaluation Questions

- Does the carrier's initial EFB training include evaluation of both knowledge and skill requirements?
- Does the carrier's recurrent or continuing qualification training include evaluations of proficiency with the EFB during all appropriate evaluation gates?

2.3.6 Fidelity of EFB Training Device

Training/Procedures Recommendation

> - The level of EFB software and hardware fidelity for training should match or exceed the level needed to achieve the specific training goals.

Problem Statements

The issue of what constitutes sufficient simulator fidelity for training and evaluation purposes has been of interest to the air carrier community for many years. Historically, full fidelity was assumed to be the ideal. Now, it is known that the required degree of fidelity depends upon the specific training goals. Training device fidelity is also an issue for EFBs and should also be driven by training requirements.

Fidelity requirements should be driven by the training requirements during each stage of the training process. A clear definition of the training goals should drive decisions concerning the required level of fidelity. Depending upon the training goals, different levels of fidelity may be required of the software and hardware.

Examples

The cost of a fully ruggedized EFB that meets all certification requirements for use on the flight deck may be prohibitive when the intended use of the device is for training purposes. In addition, the availability of high fidelity EFBs may be limited. Instead, users may be able to meet many knowledge requirements by substituting a standard laptop computer that runs the EFB software. The software itself should be identical to that used on the flight deck but the hardware may not need to be.

Training the user on the procedures for operating the device will require that the hardware interface be identical to the device used in flight, but the training device need not be ruggedized.

If the training is on EFB functions that are linked to phase of flight, such as integrated ECLs, the EFB may need to be modified to allow the device to be reset to different flight phases or repeat maneuvers as necessary in a non-real time manner. If this capability is not implemented, negative training may result from the EFB function being out of sync with the simulator phase of flight, causing the flight crew to intervene in unrealistic ways.

Evaluation Questions

- Does the EFB training device that is used during each phase of training provide the required degree of fidelity?
- Does it simulate the important aspects of the task conditions, and are those aspects of the tasks that are not simulated unimportant?

General EFB System 31

2.3.7 User Feedback

Training/Procedures Recommendations

> - A formal process for gathering feedback from all personnel whose jobs are impacted by the EFB should be implemented for Part 121 and Part 135 operations. Such a process is recommended for use during design, installation, modifications, or improvements to procedures and/or training. It may be possible to integrate the process for reporting EFB feedback within existing reporting systems.
> - User feedback obtained from this process should be provided to the manufacturer (as appropriate) so that design improvements can be made.

Equipment and Training/Procedures Suggestions

> - A representative sample of users (e.g., including line pilots, technical pilots, and training captains) should be involved in the development and evaluation of the EFB and its applications.
> - Evaluation at early stages of the lifecycle could involve individual interviews, group discussions and use of paper based simulations. As the development proceeds, increasingly realistic simulations of the system could be provided, for example using personal/desktop computer simulations. In the later stages of the development, before final operational use, feedback could be collected from aircraft simulator sessions.
> - The introduction of the EFB into a fleet may benefit from small-group tryouts of the EFB, operating procedures, documentation, and training.

Problem Statement

Developing and introducing a new piece of equipment into the flight deck requires changes to procedures, documentation, and training programs. Users are an important source of feedback throughout both development and initial deployment. The feedback should be directed to a group whose purpose is to track issues and design features, so that requests for modifications are explored and implemented as necessary.

Examples

A small-group tryout prior to full introduction into a fleet can be an effective way of evaluating changes to procedures, documents, and training—particularly on an early system prototype prior to finalizing the design. Rapid feedback can be obtained and the investment in staff training is minimized.

Once the EFB has been introduced, a formal process for accessing feedback from users can provide valuable information as users become experienced with the device. Pilots, check airmen, instructors, dispatchers, and other personnel should be encouraged to submit their opinions and suggestions for improvements concerning procedures, techniques, documentation, problems occurring on the line, and training. Each submission should receive a formal response from an appropriate manager.

Advanced Qualification Program (AQP) carriers can use their data analysis methodology to gather evaluation information from all evaluation gates, including line checks and Line Oriented Evaluations (LOEs), to assess how well pilots are doing with the EFB.

Evaluation Questions

- Were representative users involved in the design of the EFB?
- Will the EFB be introduced using small-group tryouts? Does the tryout group include representatives from all user groups, including pilots of glass-cockpits and non-glass cockpits, maintenance personnel, and others noted above?
- Is there a formal process for gathering feedback about the EFB and its support? Will feedback from this process be sent to the equipment manufacturer?

2.4 Software Considerations

2.4.1 User Interface—General Design

Equipment Recommendations

> - The EFB user interface should provide a consistent user interface by providing a common set of controls (e.g., buttons) and graphic elements (e.g., icons, windows, and menus) across all applications.
> - Controls that are used for different purposes should be visually distinct from one another (e.g., in shape or size).
> - Functional properties of graphic elements and controls should follow standard personal computer conventions, except where there is evidence that these conventions are inappropriate for use in the flight deck environment.
> - Within menus (soft key, pull-down, or any other type), functions should be accessible in proportion to their frequency of use and criticality to the mission of the aircraft.

Equipment Suggestions

> - The user interface should be designed in accordance with appropriate industry guidance materials (e.g., style guides for industry operating systems, or guidance from the Society of Automotive Engineers (SAE) and RTCA).
> - Whenever possible, controls and graphic elements for the user interface should be derived from a common software library, to ensure a high degree of consistency.

Equipment Design Tradeoff(s)

> During the initial design of EFB hardware, designers should consider what functions would be better handled by hardware controls (i.e., physical knobs and buttons), and what functions would be better handled by software controls. Functions that need to be accessible quickly or frequently, regardless of what application is active on the EFB (e.g., brightness controls), are more suited to hardware controls. Functions that are specific to an application might be better suited to software controls.

Problem Statement

The user interface is often the main aspect of a device that affects a user's ability to use a system effectively and efficiently. A device could have many functions, but if the user interface works poorly, users may choose not to use the system, or they may make errors (noticed or unnoticed) that increase workload and have other negative consequences.

User interfaces that work smoothly, and without concerted effort on the user's part, are those that work the way users expect them to, are internally highly consistent, and have anticipated the user's commonly accessed functionality appropriately.

An often overlooked, but perhaps the most important, aspect of usability is the underlying structure of the user interface. If the functions, the logic used to access and operate those functions, and the means of navigating through the user interface are consistent with existing user knowledge, users will be more likely to learn how to use the product with minimal training and find it intuitive. If the functions, functional logic, and navigation logic are inconsistent with user expectations, arbitrary, or driven by technological considerations, users will likely find the product hard to use.

Consistency is important, not just within the product, but between the product and the familiar world, the task environment, cultural conventions and expectations, industry and user interface standards, and other similar products with which the user may be familiar.

Examples

Functions that are common across applications should be performed in a consistent manner. For example, there should be high level similarities between the tasks of accessing a reference document, accessing a chart, and accessing a checklist.

Another example of a common function is selection of an airport identifier. Every time an airport identifier is entered, regardless of the application or context, the same procedure(s) should be available.

Controls that are similar in functionality (e.g., those that perform actions) should be graphically similar to each other (e.g., in shape and color).

Users find the Windows and Macintosh graphical user interfaces easy to use because the interaction logic is based on a real-world counterpart that they are already familiar with: the world of paper documents, file folders, desktops, garbage cans, and so forth. It is the desktop metaphor, not just the graphics, that makes these interfaces intuitive. Where appropriate, use of a metaphor may help make an application easy to use.

Evaluation Questions

- Is the user interface highly internally consistent? Are there standard ways to perform common actions? Are a common set of controls and graphical elements used?
- Are the user interface, functions, and functional and navigation logic consistent with user expectations, based on their prior experience and knowledge?
- Are common actions and time-critical functions easy to access?

2.4.2 Application Compatibility and Style Guides

Equipment Recommendations

- All applications on the EFB should be designed in conformance with a style guide. If available, the style guide should be customized for the specific aircraft, otherwise the style guide could be for that specific EFB system. The practices documented in the style guides should be evaluated for suitability in an aviation environment.

  NOTE: Style guides for EFBs should also be consistent with other flight deck style guides as well as industry standard operating systems, such as Windows™ (see list in Reference section).

- There should be an internally consistent user interface across EFB applications, including but not limited to data entry methods, the use of color and other formatting (e.g., use of underlining), and symbology. (AC-120-76A, Section 10.b (1))

- If available, the help facility should be standardized across the EFB applications.

- Soft key labels and drop down menus should also be consistent across applications. (See Section 2.5.2 Hardware Controls for more information on the design of soft key labels)

- Specific common actions that are allowed on multiple applications (e.g., launching or exiting an application, or selecting a hyperlink) should be performed in the same manner.

- If an industry style guide is not available, developers of EFB hardware and operating systems should write a user interface style guide for application developers to explain the user interface principles and conventions for that system. These style guides should include standard practices for performing common actions (e.g., opening and closing documents, selecting and editing text, or printing) and standard user interface elements (e.g., windows, menus, dialog boxes, and system alerts). Conventions for the behaviors of the input mechanism (e.g., single- or double-clicking), use of color, and icons/graphic elements, standard shortcut conventions, and navigation methods should also be documented. Portions of the style guide may be of use to the end user during training (e.g., see 2.3.2)

  NOTE: For guidance on consistency with the flight deck and other systems, see 2.1.3 and 2.1.4.

Problem Statement

To date, EFB prototypes have all used software customized for them by a single vendor. But in the near future, customers may purchase EFB software from different vendors, much as today one can buy software written by different companies that runs on a single desktop personal computer. If the applications are not written using the same types of graphical user interface elements and conventions, users will take longer to learn the individual software applications. In addition, users will be more likely to make errors and become frustrated with all of the applications, not just the ones that are different from the standard applications provided by the original system developers.

Manufacturers who are constructing their own EFB operating system and library of graphical controls and elements should keep in mind that internal consistency of the system and external consistency with familiar conventions are critical to user acceptance and system usability. Style guides can greatly improve the internal consistency of applications.

Examples

Software style guides are common in the personal computer industry (see reference list). Manufacturers of operating systems specify the conventions they use in these style guides so that application developers can build consistent graphical user interfaces. Another advantage of style guides is that they can shorten the time to develop applications by giving software developers guidance on the correct user interface design methods.

In the context of electronic documents, there should be consistency in terms of text color and formatting conventions and the use and placement of windows and/or frames (if implemented). Other areas where conventions may need to be established include the use of shortcut keys, abbreviations and terminology, and use of graphical elements such as command buttons and scroll bars.

To distinguish consistently between visited and unvisited hyperlinks, unvisited hyperlinks could be identified as underlined text of one color, and visited hyperlinks could be underlined text of a different color.

Heading styles and formats could be used to highlight specific types of text, such as captions, special notes or caution messages.

Evaluation Questions

- Does the EFB software conform to any existing style guide for that system?
- Did the manufacturer of EFB hardware write their own style guide?
- Is the style guide consistent with established user interface conventions for similar systems?

2.4.3 General Use of Colors

Equipment Requirements

> - Red and amber should be used only to indicate *warning* and *caution* level conditions respectively. (AC 120-76A, 10.d (1), see also 2.4.8 on Alerts and Reminders.)
>
>   NOTE: If there is a regulatory or TSO requirements that conflicts with the recommendations above, those supersede this guidance. (AC 120-76A, 10.d (1))
>
> - Color should not be used as the *sole* means of coding important differences between information; instead color should be used redundantly with other coding (e.g., shape or size of the object, or text labels).
> - Any color-coding scheme used should be interpretable easily and accurately.

Equipment Recommendations

> - Each color used in a color-coding scheme should be associated with only one meaning.
> - In general, no more than six colors should have assigned meanings in a color-coding scheme. (SAE ARP 4032)
>
>   NOTE: The use of more than six colors is not precluded, but careful testing should be undertaken to ensure that the use of a larger number of colors with assigned meanings *improves* task performance.
>
>   NOTE: More than six colors can be used as long as the other colors do not need to be identified, but only recognized as being different from the other colors. (Cardosi and Hannon, 1999)
>
> - EFB color-coding schemes should not conflict with flight deck standards for that particular aircraft. For example, green is generally assigned to indicate a safe condition, and magenta may also be reserved to indicate either flight data source or active flight path targets in some flight decks. These colors should be avoided for any other purpose. (See also 2.1.3 and AC 25.11)
> - If colors are customizable, there should be an easy way to return to a default color-coding scheme. This default scheme may be specified by the manufacturer, the Part 91 user, or an administrator for Part 121 and 135 operations.

Installation Recommendations

> - Colors on the EFB should be discriminable by the typical user under the variety of lighting conditions expected in a flight deck from a nominal reference design eye point.
> - Each EFB should be evaluated for the effects of its positioning on how well colors can be discriminated. The ability to discriminate colors may also be affected by how the unit is positioned with respect to the user, e.g., whether the unit is used in the portrait or landscape mode. Cross-cockpit viewing may not be acceptable if the colors wash out significantly.

Training/Procedures Recommendations

> - For Part 121 and 135 operations, default colors used in color-coding schemes should be customizable only by an appropriately authorized administrator, not by the end user flight crew. The administrator should also ensure that the EFB color scheme is compatible with other color conventions used in the flight deck.
> - Users should receive training on the standard color conventions used on the EFB.

General EFB System   37

Equipment Suggestions

> - Upon request, the EFB should display a legend describing the meanings associated with different colors. Access to a legend of colors is especially important if the colors are user customizable.
> - Colors used in a coding scheme should be widely separated from a perceptual point of view. In particular, the following color pairs are perceptually very similar and therefore confusable: red/magenta, magenta/purple, yellow/amber, and cyan/green. A color-coding scheme should not employ both members of a pair because they are difficult to distinguish reliably.
> - Large areas filled with saturated colors (e.g., rectangles filled with pure red or blue) may cause eyestrain and/or afterimages and should be avoided.
> - Pure blue should not be used for small symbols, text, fine lines, or as a background color. (Cardosi and Hannon, 1999)

Equipment Design Tradeoff(s)

> There may be special considerations for EFBs to be used with other devices, such as head-up displays or night vision goggles, where additional lighting issues need to be considered.
>
> The display brightness setting may affect color discriminability.

Problem Statement

Information that is displayed in colors that are not discriminable will cause pilot distraction at the least (as the pilot attempts to position the display for better legibility) and potentially more harmful consequences if important information is misread, or not read at all. Color discrimination can be compromised by a variety of factors including lighting conditions, viewing angle, size of the object, and display technology, quality, and calibration. Additionally, color perception varies across individuals. As the eye ages, its ability to focus on red objects or differentiate between blue and green is reduced. In particular, pure blue is problematic for observers over age 50.

If color coding is used to indicate grouping, and if an individual EFB unit may be used by more than one person (as in Part 121 or 135 operations), it is important that the color-coding scheme be standardized by an administrator. Otherwise, one user may configure a color to an opposite or unfamiliar meaning, confusing another user. In Part 91 operations, a greater degree of individual end-user configuration may be supported, even though this is not recommended.

Some colors have strong associations in general, such as red. Some common associations for red are danger, emergency, failure, stop, no-go, and fire/hot (CAP 708, see References). Also, there are strong flight deck conventions for the use of red (see 2.4.8 on Alerts and Reminders). As a result, any use of red should be considered carefully so that it does not conflict with pilot expectations. Also, in order for red to retain its distinctiveness in the flight deck, it is important to limit its use to the highest priority situations, such as when pilot actions are time-critical. If red is used too broadly, pilots may not quickly be able recognize situations where their actions *are* time-critical. Where there is a conflict between this intended meaning for red, and other strongly held pilot expectations (e.g., the use of red to denote high intensity precipitation, which may be far from the aircraft and therefore does not represent a situation requiring immediate crew action), the use of red may be allowed through additional FAA-approved guidance material, which would supersede the guidance in this document.

Examples

One way to ensure that colors are used redundantly with other cues is to design the system for a monochrome display first, and then add color afterwards. (CAP 708)

Different font sizes may be used along with color to discriminate between different types of text data. For example, a large white font may denote the active checklist item and a small green font style could represent a completed item. If the font encodes important information, all of the fonts on the screen should be legible and easily discriminated from one another without any screen adjustments and in all lighting conditions.

Some display technologies are meant to be viewed only in one orientation, e.g., either portrait or landscape. If the software requires or allows screen content to be rotated to different orientations, the screen should be just as readable and the colors the same when the device is rotated. For example, if a chart could be viewed in either portrait and landscape mode, there should be no difference in readability, brightness, color rendering, and so forth when the device is viewed in either orientation.

Unless monitors are calibrated appropriately, the appearance of a color defined by Red-Green-Blue (RGB) or Hue-Saturation-Lightness (HSL) values will vary. In order to specify a color accurately, Commision Internationale de l'Eclairage (CIE) color coordinates are used, and monitors are calibrated to ensure that they are displaying the defined color correctly.

Highly saturated color may cause afterimages. For example, when one stares at a solid blue object for a period of time, upon looking away, one may see a yellow afterimage in its place.

Evaluation Questions
- Are red and amber used, and if so, is it used appropriately?
- Are colors that convey grouping used redundantly with other cues? Is it possible to understand all the information being conveyed when the screen is viewed in monochrome?
- If colors are used, are they all discriminable under the various conditions that the EFB will be used in?
- If colors can be customized, is it possible to define color schemes that would violate flight deck color standards?

General EFB System 39

2.4.4 Graphical Icons

Equipment Recommendations

> NOTE: Graphical icons are software-implemented controls that are represented on the screen by pictures of limited size and resolution.
>
> - Graphical icons should be accompanied by brief text labels. (See also 2.5.2.)
>
>   NOTE: 14 CFR 23.1555, 25.1555, and 27.1555 require labels for all controls unless their function is obvious.
>
> - Graphical icons should be designed carefully to minimize any necessary training, and to maximize the intuitiveness of the icon for cross-cultural use.

Training/Procedures Recommendation

> - Users should receive training on the meaning of graphical icons used on the EFB to ensure that the icons are understood.

Equipment Suggestions

> - Users should be able to access text help information to explain meaning of graphical icons in more detail than the text label alone.
> - Using the same graphical icons as other flight deck systems can reduce training requirements and help to prevent error, especially in high workload phases of flight or during abnormal or emergency situations.

Problem Statement

Graphical icons could be used to access commands from graphical menus/toolbars, or they may represent files and other system objects. Icons can be especially useful for expert users, especially if the graphics have obvious meanings.

One of the common problems in icon design, however, is that the intended meaning of the icon is not immediately clear. Untrained or cross-cultural users may have significant trouble understanding the intended meaning of the icon, especially since icons are typically small and of limited graphical resolution. There are many computer applications with icons whose forms are not obviously tied to the functions they represent.

The use of graphical icons in and of themselves does not ensure usability. When the icon is too abstract, the user has to learn and remember its meaning, and the icon is no longer useful as a memory aid. Even trained users may forget the meaning of an icon if it is not used frequently, or if they are in stressful conditions.

Because of their potentially critical role, it is important that users are trained on the interpretation of icons and it is important that the actual graphical image be a redundant, not sole, means of representing an object or command. Text information about that object or command should also be available.

Examples

Achieving familiarity with all EFB icons should be an important goal of initial EFB training. In the event that the user forgets an icon's meaning, the EFB should provide an easy means for finding that information. One approach is to provide a textual label for the icon if the cursor lingers on that icon. Another approach is to provide a "legend" of all icons and their meaning that is available in a central location. A third approach would be to limit the number of icons that are used, so that there are only a few to learn. Ideally, the intended meanings of all icons should be so obvious that minimal training and no memorization or use of references are necessary.

Standard icons have been developed for road signs (e.g., stop signs, pedestrian crossing, yield). These icons are documented in booklet from the USDOT Federal Highway Administration titled *Road User Guide for*

North America (DOT FHWA-SA-99-020). The icons in the booklet are well designed in that their meanings are intuitive across cultures. Common signs are especially similar across cultures.

Evaluation Questions
- Does the initial EFB training adequately address icon meanings?
- Are icons designed for ease of interpretation to minimize the need for training?
- Does the EFB provide easy access to help information that explains each icon's meaning?
- Does the EFB use icons in a way that is consistent with other flight deck systems?

General EFB System   41

2.4.5 Multi-Tasking

Equipment Recommendations

> NOTE: These recommendations apply if the EFB is able to run more than one application at the same time.
> - The user should always be able to identify the currently active application easily.
> - The user should be able to:
>
>   (a) select which of the open applications is currently active
>
>   (b) switch between any of the available applications easily, even when there are pending tasks in the foreground application
>
> - When the user returns to an application that was running in the background, it should appear in the same state as when the user left that application, other than any differences associated with the progress or completion of processing performed in the background.
> - The responsiveness of any individual application (e.g., time to process user input or complete computations) should not suffer if all other supported applications are running at the same time.
> - To exit applications that have pending activities, the user should deal with these activities, either by completing them, or by acknowledging that they are incomplete.
> - Some EFBs will support applications that are not directly related to flight tasks (e.g., word processor or game). In order to discourage inappropriate use of non-flight-related functions, the system should, at a minimum, warn the user that these applications are not recommended for use in flight, and ask for confirmation to proceed when they are launched.

Equipment Suggestions

> - It may be useful to launch some applications automatically upon system startup. Switching between applications that are already open may be faster than opening an application at the time of the requested switch.
> - For each application running at the same time on an EFB, it may be useful to have summary information available in the foreground, without requiring the user to activate that application. The operating system should support the ability to view basic information about an application without activating that application.

Problem Statement

Multi-tasking may be implemented and used operationally in different ways. The main priority to keep in mind is that the user has to be able to keep track of which application is currently active easily in order to prevent confusion. Sometimes the user may need to switch between multiple active applications quickly. If switching is cumbersome, time-consuming, or error prone, the pilot's workload will increase. In order to avoid switching between applications often, it may be useful to be able to view some basic information without having to activate the background application.

Examples

One way to manage multiple applications is from a soft-key "main" menu that has a list of all available applications. Users should be able to get to this main menu with a single key press from any point in the software. The main menu should appear on the default starting screen.

Some applications, such as an ECL or chart display, could start automatically when the EFB is turned on since they may be useful immediately. These applications should also close automatically when the EFB is turned off. The user should still be able to close them manually, after seeing a message to finish pending tasks and user confirmation. For example, if the user attempts to close the ECL, and there are one or more incomplete checklists, the user should deal with checklists before exiting the ECL.

For an ECL that is running in the background, it may be useful for the user to know the name of the currently active checklist, if any, in the foreground. This technique is especially helpful when the background application has time-dependent information (e.g., traffic information). For traffic information, for example, a running count of the number of other aircraft within a user-specified distance, such as 10 miles, may be useful to have in the foreground.

Evaluation Questions

- Can the user identify the current application easily? Can the user switch between applications easily?
- Is there an extra confirmation step required to get to any applications that are not flight related?

General EFB System 43

2.4.6 Responsiveness

Equipment Requirements

> - The system should provide feedback to the user when a user input is processed. Alphanumeric inputs should be drawn on the display within 0.2 seconds. (SAE ARP 4791)
> - If the system is busy with internal tasks that preclude immediate processing of user input for longer than 0.5 seconds (e.g., calculation, self test, or data refresh), a "system busy" indicator (e.g., clock icon) should be displayed to show the user that the system is occupied and will not process inputs immediately. (SAE ARP 4791)

Equipment Recommendations

> - System-busy indicators should be common across all hosted EFB applications.
> - The type of feedback should be appropriate for the type of user input.
> - If an internal task takes more than a few seconds to complete and user entries are not processed immediately while the task is in progress, a progress indicator should be displayed to show the user how much of the internal task is complete so that he/she has a sense of when the full task will be completed and when processing of user input will resume.
> - User entries that are made while the system is occupied should be stored and processed as soon as resources are available.

Equipment Suggestion

> - A busy indicator can be drawn *whenever* the application is occupied. If shown for only a short time, the busy indicator is not noticeable; if shown for a longer time, it will be noticed.

Equipment Design Tradeoff(s)

> Responsiveness may be more important for functions used in high workload phases of flight, but even less important applications should be reasonably responsive. End-user (pilot) input may be needed to determine what is "reasonable."
>
> The responsiveness of an application may be more variable, and slower overall, if files are stored remotely and accessed via network or data link.

Problem Statement

Immediate user feedback should be provided so that the user is aware that the system has received and accepted his/her input. Without feedback, the user may try to re-enter inputs multiple times in a short interval, and he/she will become confused when the system eventually acts upon the multiple inputs. If user entries are not being accepted, this should be indicated so that the user does not attempt to make entries that are then discarded by the system. Excessive delays in access to requested information may have significant negative effects, especially during high workload phases of flight. System-busy indicators and appropriate user feedback can help users to understand that their input has been received and will be processed as soon as possible.

Examples

Examples of feedback for user inputs include:
- If the user enters a character, the character is drawn on the screen.
- If the user manipulates the cursor pointing device, its location is updated and redrawn.
- If the user selects an action from a button, that action is started.

Some commons shapes for busy indicators include clocks, hourglasses, or spinning balls/dials. Progress indicators may consist of graphical bars that are shaded in proportion to the degree that the task is complete. Text progress indicators (shown in percent complete or seconds to complete) are also acceptable.

One common problem with progress indicators is that they are often misleading. In some applications, the estimate of progress or time remaining may be based on the number of software modules remaining to be accessed, rather than how much time the entire process will take. Consequently, some applications' progress indicators may, for example, take a very long time to achieve 50% completion and a very short time for the rest, or vice versa. If progress indicators do not accurately represent the true time remaining or the percent of total time complete, users may learn to disregard them.

While access to flight manuals is not always a time-critical task, documents should be loaded quickly when requested. Electronic charts may need to be available more quickly than reference information, because they may be needed during high workload phases of flight, such as while on approach to landing.

Evaluation Questions

- Is feedback provided for all types of user inputs within an appropriate time? Are busy indicators displayed when processing of user input is delayed?
- Are progress indicators displayed for tasks that take a significant amount of time? Are these indicators accurate and clear? Do they represent progress in a way that is useful to the pilot?

General EFB System 45

2.4.7 Anchor Locations

Equipment Recommendations

> - There should be an anchor location from which the user can move between EFB applications if more than one application is supported.
> - Each application supported on the EFB should have its own main/anchor page, from which functions within that application are accessible.
> - It should be easy for the user to move from any location in the EFB to an anchor location, such as the main menu, or home page for that application.

Problem Statement

In order to be useful to the pilot, the information on an EFB needs to be quickly and easily accessible. In order to be accessible, the pilot should know how to get to any desired data quickly, starting from any other point in the EFB. One of the standard ways to help the user orient him- or herself is to provide an anchor location such as a top-level main "menu", or a "home page." Anchor pages are appropriate within each application. There should also be an anchor page from which different EFB applications can be accessed. When anchor locations are easy to access, the user can jump from one place in the EFB to another quickly and easily, while staying oriented. A common example of this is the "home" page of a web site.

If there are no anchor locations, the user is more likely to get disoriented and have trouble moving from one place to another in the EFB. This results in increased workload, distraction, and head-down time in order to navigate the user interface.

Examples

In a full graphical user interface, the "desktop" could be the main anchor location. It would be easy to return to the desktop from anywhere in the EFB functionality. From the desktop, the user could determine what EFB functions were running, and switch to any other function.

In a soft-key menu-based user interface, a top-level menu could be the main anchor location. The user would be able to go to that top-level menu by pushing a single, dedicated button at any time. Individual EFB applications, such as an electronic document viewer or a chart display, could have their own anchor pages, from which the user could access any of the functionality of that particular application, such as opening a document, or viewing the graphical chart.

Evaluation Questions

- Where are the anchor locations in the EFB software user interface? Are they obvious and easy to get to from any position in the software?
- Are there anchor pages within the applications? Is there anchor page to go between applications?

2.4.8 Alerts and Reminders

Equipment Requirement

- EFB alerts and reminders should meet 14 CFR Part 23.1322, 25.1322, 27.1322, or 29.1322 as appropriate for the intended aircraft. While the regulations refer to discrete annunciation lights, the intent should be generalized to extend to the use of colors on displays and controls. (See also 2.4.3) (AC 120-76A, 10.d (1))

- Red should be used only to indicate a *warning* level condition (i.e., a condition that may require immediate action). (AC 120-76A, 10.d (1))

- Amber should be used only to indicate a *caution* level condition (i.e., a condition that may need corrective action in the near future). (AC 120-76A, 10.d (1))

- Colors other than red and amber may be used for items other than warnings or cautions, providing that the colors used differ sufficiently from the colors prescribed to avoid possible confusion. (AC 120-76A, 10.d(1))

NOTE: If there is a regulatory or TSO requirements that conflicts with the recommendations above, those supersede this guidance. (AC 120-76A, 10.d (1))

Equipment Recommendations

- EFB alerts and reminders should be consistent with the FAA standards for electronic displays in AC 25-11 and 14 CFR Part 23.1311a. International recommended guidance can be found in publications of the Joint Aviation Authorities (JAA) including AMJ 25-11 and AMJ 25.1322 on warnings and cautions. See also SAE ARP 4102/4 and the FAA HFDS for further guidance.

- EFB alerts and reminders should be integrated with (or compatible with) presentation of other flight deck system alerts. (AC 1 20-76A, 10.d (1))

- Messages should be prioritized and the message prioritization scheme should be documented and evaluated. (AC 1 20-76A, 10.d(1) and Section 10.d (2))

- Text or symbols that flash too quickly and brightly can grab and hold the pilot's attention so strongly that they are distracting. Such strong attention-getting techniques should be avoided in any EFB application. (AC 1 20-76A, 10.d (1))

- During high workload phases of flight,

  (a) Required flight information should be continuously presented without uncommanded overlays, pop-ups, or pre-emptive messages, except those indicating the failure or degradation of the current EFB application. (AC 1 20-76A, 10.d (1))

  (b) EFB messages, both visual and auditory, should be inhibited, except those indicating the failure or degradation of the current EFB application. (AC 1 20-76A, 10.d (1))

Installation and Equipment Design Tradeoff(s)

Designers should consider whether the alert is necessary and where it should be displayed, particularly for EFBs that communicate with flight deck systems. Alerts could be displayed on the EFB, or on a general flight deck message display. If some messages are shown in one place, and others are shown elsewhere, there should be a clear rationale, from the crew's perspective, as to what types of messages are shown in each location.

Any use of audio alerts should be assessed in terms of confusion with other audio alerts.

General EFB System   47

Problem Statement

Because EFBs may be integrated with flight deck systems, and/or present information that is critical to flight safety, it is important to integrate flight-critical EFB alerts/reminders into the overall flight deck alert/reminder philosophy. Some of the factors to consider in the display of alerts include (a) overall necessity, (b) time-criticality, and (c) priority level within the overall context of the flight.

Because the flight deck environment places tremendous demands on the pilot visual system, audio alerts are often used. Audio is already used on the flight deck by a number of systems. Ten or more unique audio sounds, together with dozens of vocal warnings, are not uncommon on advanced automation aircraft. An audio warning system as part of an EFB should be assessed in terms of possible confusion with other systems, ease of control, and training requirements.

Example

Messages that are time-critical and relevant to the safety of flight, although they may be generated by an EFB, should be displayed on a general alerting system. Messages specific to the EFB application (e.g., "value out of bounds" or system status messages) should be displayed on the EFB.

It may also be appropriate to display messages with different amounts of detail both on the EFB and on the general alerting display. For example, the general alert display could show generic text such as "EFB Message" and the specific message would be shown on the EFB.

See the Terrain Awareness and Warning System (TAWS) TSO C151a for a sample alert prioritization scheme.

If the EFB is deployed during takeoff, and it has any audible alerts, it should be connected to the avionics in such a way that it respects the alert suppression logic used throughout the flight deck for takeoff, the initial stage of climb, and the final stage of descent and landing.

Evaluation Questions

- For installed systems, do EFB alerts and reminders meet the requirements in the appropriate regulations (listed above)?
- What is the philosophy for alerts/reminders?
- Are strong attention-getting flashing symbols avoided?
- Are EFB messages inhibited during high workload phases of flight unless they pertain to the failure or degradation of the current EFB application?

2.4.9 Display of System Status

Equipment Recommendations

> - If an application is fully or partially disabled, or is not visible or accessible to the user due to a failure, this loss of function should be clearly indicated to the user with a positive indicator. That is, lack of an indication is not sufficient to declare a failure condition. (AC 120-76A, Section 10.d (2))
>
> - The immediacy of the status annunciation should be appropriate to the function that is lost or disabled. For some functions, an immediate status annunciation (i.e., an active interruption) is appropriate. For other functions, a passive status indicator that appears when the user attempts to access that function is more appropriate. (AC 120-76A, Section 10.d (2))

Problem Statement

There are many reasons why systems fail to operate as expected. For example, functions that require external data may fail when that data are not received, producing a partial failure. A total failure may occur if there is a hardware fault.

The user should receive a positive indication of any system failures. A positive indicator of failure (e.g., a light or message that appears upon a failure condition) is clearer and more noticeable than a negative indicator (e.g., a light or message that turns off.) The immediacy of the status annunciation should be matched with the characteristics (e.g., importance and time-criticality) of the function that is lost or disabled.

Without a clear indication of the failure, the user may make decisions based on outdated, incorrect, or incomplete information.

Examples

Access to electronic charts may be lost. However, if paper backups are required to be in use, the loss of electronic charts may not require an audible (high-immediacy) alarm.

If the EFB application is integrated with other flight deck systems and this connection fails, the user should be alerted immediately to this failure. For example, an EFB electronic checklist may be designed to bring up a non-normal condition checklist upon encountering such a condition. However, if the EFB has lost its connection, it may not know of the non-normal condition and the pilot may miss an associated checklist because he/she presumed that the checklist function was working correctly when the non-normal condition began.

As another example, if the EFB provides real-time information, such as weather or air traffic clearances via some type of data link system, the user should be notified if there is a problem with the data link that precludes normal display of the data. If the link is down completely, and there are no data to display, this should be distinguished from the case where there is a blank screen because, for example, there is no traffic or precipitation in the selected region. If the link is operational in a degraded mode (e.g., the data rate is half of the normal rate so that the data are refreshed less often), this should also be brought to the crew's attention. (Note: For data link services, this requirement to notify the crew of system errors should be consistent with the appropriate Minimum Operational Performance Standards for that service, available through RTCA, and with any related FAA approved guidance materials, e.g., AC 20-140 on Aircraft Data Communication Systems.)

Non-essential applications such as e-mail connectivity and administrative reports may require an error message when the user actually attempts to access these functions, rather than an immediate status annunciation when the failure occurs.

Evaluation Questions

- Are partial or full failures of the EFB clearly annunciated with positive indicators?
- Is the immediacy of the failure annunciation appropriate to the function that is lost or disabled?

General EFB System _____ 49

2.4.10 Legibility of Text—Characters

Equipment Recommendations

> - The EFB should use a highly legible typeface that enables the user to quickly and accurately identify each character. In particular, the HFDS recommends:
>
>   (a) Upper case text should be used sparingly. Upper case text is appropriate for single words, but should be avoided for continuous text. (HFDS 8.2.5.8.2)
>
>   (b) For continuous text (e.g., sentences and paragraphs), use mixed upper and lower case characters. (HFDS 8.2.5.8.4)
>
>   (c) Use serif fonts for continuous text if the resolution is high enough not to distort the serifs (small cross strokes at the end of the main stroke of the letter). (HFDS 8.2.5.7.5)
>
>   (d) Sans serif fonts should be used for small text and low resolution displays. (HFDS 8.2.5.7.6)
>
>   (e) For optimum legibility, character contrast should be between 6:1 and 10:1. Lower contrasts may diminish legibility, and higher contrasts may case visual discomfort (HFDS 8.2.5.6.12)
>
>   (f) Characters stroke width should be 10 to 12% of character height. (HFDS 8.2.5.6.14)
>
> - Individual characters should not be easily confused with other characters. Characters that are most likely to be confused are:
>
>   —P and R
>   —B, D, and E
>   —G, O, and C
>   —l (the letter) and 1 (the number)
>   —Z and 2
>
> - Use of slanting or italic text should be avoided.

Problem Statement

In order for information to be quickly and accurately understood by the user, each typeface that is used by the EFB should be highly legible. Legibility is affected by such factors as the shape of each character, the width of the strokes that form each character, and contrast between the character and the background. Italic and upper case text are generally more difficult to read (i.e., they are more prone to being misread, and take longer to process even when read correctly) than plain text.

Examples

Sans serif typefaces do not use small horizontal strokes at the top or bottom of characters (e.g., "h" or "y"). Serif typefaces do have small horizontal strokes at the top or bottom of characters (e.g., "h" or "y"). Sans serif typefaces are typically more legible than serif typefaces on low-resolution displays. If a serif typeface is used, a high screen resolution is necessary to achieve comparable legibility.

Upper case text is more difficult to read and, therefore, should be used sparingly. It should not be used for emphasis. Slanted or italic text should be avoided for the same reason. To convey emphasis, one suggestion is to increase font size or place highlighting characters (e.g., asterisks) on either side of the emphasized text.

Evaluation Questions

- Are individual characters easily recognized for each typeface that is used?
- Does the typeface use strokes with sufficient and constant width to enable each character to stand out against the screen background?
- Are upper case and italic text avoided?

2.4.11 Legibility of Text—Typeface Size and Width

Equipment Requirement

- The EFB should use a typeface size that is appropriate for the viewing conditions (e.g., viewing distance and lighting conditions) and the criticality of the text.

Equipment Recommendations

- The FAA HFDS recommends the following:

  (a) The minimum character height should be 16 minutes of arc (5 millirad). For practical purposes, this requires a minimum typeface height of 1/200 of the viewing distance. (HFDS 8.2.5.6.6)

  (b) The preferred character height is 20 to 22 minutes of arc (approximately 6 millirad). For practical purposes, this translates into a typeface height of 1/167 of the viewing distance. (HFDS 8.2.5.6.5)

  (c) The ratio of character height to width should be (HFDS 8.2.5.6.10):
  —At least 1:0.7 to 1:0.9 for equally spaced characters and when lines of 80 or fewer characters are used.
  —At least 1:0.5 if more than 80 characters per line are used.
  —As much as 1:1 for inherently wide characters such as "M" and "W" when proportionally spaced characters are used.

  If these guidelines are not met, there should be a sound basis for deviation.

- A larger typeface size should be used for text that can be expected to be read under poor viewing conditions in the flight deck (e.g., some emergency checklists).

Problem Statement

Typeface size is a critical determinant of the ease with which text can be read. The variety of lighting conditions under which the text will be read has to be considered. Equally important is the type of information being conveyed. Important information that needs to be read under potentially low-visibility conditions should be displayed using a larger height in order to ensure that the users can quickly and accurately read the information. Equally important is typeface width. Narrow characters can be more difficult to read.

If the typeface is too small to read when the EFB is in its normal use position, pilots may misread the text, or they may incur extra workload by the need either to zoom the display, or to re-position it to be at a closer viewing distance to make the text legible.

Examples

The minimum typeface size that is used should support legibility under a wide range of lighting conditions. EFB information should, therefore, be presented using a minimum typeface height of that is 1/200 of the viewing distance. For a viewing distance of approximately 30 inches (76 cm), the characters should be a minimum of 0.16 inch (3.8 mm) high. The preferred character height at this viewing distance is 0.18 inch (4.6 mm). A larger size may be required for some applications. In particular, emergency checklists that will be used under poor viewing conditions, such as checklists used for smoke-related conditions, should use a larger size.

Evaluation Questions

- Is the typeface size easily legible under normal viewing conditions?

- Is the typeface size adequate for emergency checklists and other important text that can be expected to be used under low-visibility conditions?
- If the typeface size is smaller than recommended, what level of workload is incurred by the need to zoom or re-position the display to make the text legible?

2.4.12 Legibility of Text—Spacing for Readability

Equipment Recommendations

> - Text should be spaced appropriately for ease of reading.
> - Line lengths should be appropriate for the type of text (e.g., checklists).
> - In order to make text easily readable, the FAA HFDS recommends the following:
>
>   (a) Use a horizontal spacing between characters of at least 10 percent of character height. (HFDS 8.2.5.6.1)
>
>   (b) Use spacing between words of at least one character when using equally spaced characters or the width of the capital letter "N" for proportionally spaced characters. (HFDS 8.2.5.6.2)
>
>   (c) Use a vertical spacing between lines of at least two stroke widths or 15 percent of character height, whichever is larger. Vertical spacing begins at the bottom of character descenders (that part which descends below the text line as seen in the lower-case letter "y") and ends at the top of accent marks on upper case characters. (HFDS 8.2.5.6.3)
>
>   (d) Separate paragraphs with a blank line (HFDS 8.2.5.6.4)

Problem Statement

Text displayed by the EFB should be readable as well as legible. Readability is concerned with enabling the reader to easily recognize words and keep the reader's eye from unintentionally skipping to another line of text. Readability is determined by the spacing between individual characters and words, between lines of text and paragraphs, and by the length of the line of text.

Appropriate vertical and horizontal spacing should be used between characters to facilitate reading. If the spacing of characters and words is inappropriate, the pilot may misread text and/or take longer than necessary to process the meaning of the text.

Examples

Tight spacing between characters can cause the characters to run together while loose character spacing makes the boundaries between words less detectable.

Long lines of text can cause the eye to jump to the next line. Line length is especially critical for checklists where a large gap between the challenge and response items may cause the reader to pair a response item from a different line.

Evaluation Questions

- Does the EFB use a horizontal spacing between characters and between words that clearly indicates boundaries between words but does not cause individual characters to blur with each other?
- Is there sufficient vertical spacing between lines to help the user's eye avoid skipping to the next line?
- Are the line lengths appropriate for each type of text?

General EFB System　　　　　　　　　　　　　　　　　　　　　　　　　　　　　　　　53

2.4.13 Non-Text Display Elements

Equipment Recommendations

> NOTE: Non-text display elements on an EFB could include graphical icons, or other symbols, such as those on an electronic chart.
>
> - Non-text display elements should be distinguishable based on their shape alone, without relying upon secondary cues such as color and text labels. (See also 2.4.3 and 2.4.4)
> - Non-text display elements should be designed for legibility when presented on the minimum expected display resolution when viewed from the maximal intended viewing distance.
>
> NOTE: SAE ARP 4102/7 on Electronic Display Symbology for EADI/PFD gives minimum symbol visual angles as 6 millirad for primary data, and 4 millirad for secondary and descriptive data. (See also 2.4.11 for examples of using visual angles to determine on-screen size of a character or symbol.)

Equipment Suggestions

> - Where possible, shapes of non-text display elements should be consistent with any existing paper symbol equivalents.
> - If zooming is available, symbols shapes could change as the zoom level changes (e.g., when zoomed in, a more detailed symbol could be used than when zoomed out). If this is the case, the symbols should be similar enough that users are not confused about the shape differences.

Equipment Design Tradeoff(s)

> In order to assess legibility of non-text display elements, the following factors should be considered:
>
> (a) pilot visual acuity
>
> (b) similarity to other symbols and graphics
>
> (c) smallest visual feature of the symbol that distinguishes it from other symbols and graphics
> Is the smallest identifying feature large enough to be seen easily?
>
> (d) the context in which the symbol will be shown
> Will it be one of many symbols, or stand-alone? It is easier to discriminate one symbol from another than to recognize (identify) a symbol in the absence of any context.
>
> (e) the importance of the information that the non-text item conveys
> Are there cases where a small symbol feature implies a significant difference in its operational interpretation?
>
> (f) the conditions under which the item might be viewed (e.g., position of the EFB)
> Some symbols may have fine detail that is difficult to see under degraded conditions.
>
> (g) optical qualities of the display
> Display factors affecting legibility include resolution, contrast, brightness, color, and rendering techniques such as anti-aliasing (see 2.5.3).

General EFB System

Equipment Design Tradeoff (continued)

> Symbols that are used on paper (e.g., aeronautical chart symbols) may need to be modified for better legibility on electronic displays with significantly lower resolution than paper.
>
> Standards techniques used to highlight symbols on paper charts (e.g., bolding) may not transfer well to lower resolution electronic displays. Different types of highlighting may need to be tested.

Problem Statement

Non-text display elements are essentially pictures that convey meanings. These pictures could represent symbols (e.g., elements of an electronic chart), or they could be icons that are used to access various functions. It is important that these pictures are easily recognized and distinguished at all times. Depending on the nature of the symbol, there are a range of consequences if the symbol is misinterpreted by the crew. If fine symbol detail is necessary to distinguish between symbols that represent significantly different operational impacts, and that fine detail is not easily seen, the consequences could be operationally significant. Even if misinterpretation does not occur, poorly designed non-text display elements may cause confusion, and as a result, increase crew workload.

Examples

Paper electronic charts use many fine symbols. In many cases, there are families of symbols that bear a common high-level resemblance, with fine features indicating differences. For example, there are several symbols representing airports on VFR charts. Some are filled, some are outlines only. Some airport symbols have fine notches indicating fuel availability (or lack thereof). If the filled and unfilled symbols are difficult to distinguish, or if the notches are too small to be noticeable, pilots may misinterpret these symbols.

On some paper charts, the highest obstruction in the area is a bold version of the symbol representing other obstructions. The bolding may be a good distinguishing feature on a paper chart, but it may not cause the symbol to stand out as significantly on an electronic display.

Evaluation Questions

- Are graphic objects legible and distinguishable at the intended screen resolution and from the intended viewing distance?
- Are any icons, symbols, and formatting consistent with how the same information is depicted on paper equivalents?

General EFB System 55

2.4.14 Supplemental Audio

Equipment Recommendations

> NOTE: Supplemental audio is defined as audio that is not associated with alerts and warnings. Supplemental audio could be verbal or non-verbal.
> - Use of supplemental audio while in flight should be avoided.
> - Users should be able to control the volume of supplemental audio.
> - Users should be able to turn off the supplemental audio on an EFB if desired.
> - Objects that have associated supplemental audio should be coded such that the user knows of the associated audio before it is activated.
> - Supplemental audio that is audio alone (i.e., without any visual image) should have a text description available so that the user can anticipate the content of that audio clip.
> - Users should be able to stop the supplemental audio at any time while it is in progress.

Equipment and Training/Procedures Design Tradeoff

> Supplemental audio may be useful for enhancing animation segments in a multimedia document. It could also be used for training purposes, especially if the sound is of high quality. However, operators should consider whether their policy should be to limit the use of supplemental audio in flight because the additional audio may interfere with higher priority audio information (e.g., radio communications). Supplemental audio may be difficult to hear in flight unless it is integrated into the flight deck audio system. Also, users may need training in how to use and control any supplemental audio functions.

Problem Statement

Supplemental audio is an optional EFB feature. It has the potential to significantly modernize the look and feel of the EFB. However, because it is an optional feature, users should have complete control over when (and whether) the audio is activated, and its volume. Because the flight deck has many other sources of higher priority auditory information, use of supplemental audio in flight should be avoided in general. The utility of supplemental audio may be highest for ground-based training purposes.

Examples

Video clips of training presentations, complete with supplemental audio, could be stored on the EFB. These could be accessed through a help facility.

Supplemental audio such as background music could be distracting and useless in that it does not convey any additional information to the user.

Evaluation Questions

- Is supplemental audio used in flight? If yes, is the supplemental audio audible in flight without interfering with higher priority tasks?
- What is the operator's policy regarding use of supplemental audio in flight?
- If supplemental audio is implemented, does the user have control over when, and whether, the audio is activated?

2.4.15 Ensuring Integrity of EFB Data

Equipment Requirement

> - Data that are loaded onto an EFB (e.g., databases, electronic documents, electronic charts, and even software updates) should be checked by appropriate methods to ensure that they are accurate, up-to-date, and uncorrupted prior to installation on the EFB.
>
> NOTE: AC 120-76A contains several provisions for the configuration control of data on an EFB. This AC applies to all operations other than Part 91 (excluding subpart F, to which the AC does apply). The intention of the requirement here is to make operators and users who are not subject to AC 120-76A aware of the importance of data verification. For all others, the guidance in AC 120-76A supersedes this requirement.

Equipment Recommendations

> - The EFB should validate data by checking that the current date is within the valid date range of the data.
> - The EFB should support the installation of data with an effective date in the future, so that changes may be loaded in advance.
> - When there are many sources of data for an EFB, flight crews should not be required to cross check each source for validity. Instead, the system should conduct a self-test and generate a message if any source of data is found to be out of date. The message should indicate to the flight crew where to go for further information on the self-test (e.g., to see which particular database or software is out of date).

Training/Procedures Recommendations

> - Data that are entered into the EFB by the pilot or by a remote source (e.g., an airline operating center) should be reviewed by the crew prior to use.
> - Pilots should understand where the EFB data comes from, and they should be trained to review EFB data just as critically as they review data on paper. In particular, EFB data should be judged for their applicability to current operating conditions.

Problem Statement

EFBs may provide information that is important for flight safety. This information should be based on correct, uncorrupted data. If there is any question about the integrity of the data, pilots should not be making decisions based on it. Therefore, databases should be checked prior to installation on the EFB. The checks for data integrity could involve internal software checks and/or quality control by a human.

Data that are entered into the EFB by the pilot or by a remote source (e.g., an airline operating center) should also be reviewed by the crew prior to use. If these data are not checked just prior to use, it is possible that they might have been corrupted without the crew noticing.

As an additional safeguard, pilots should be trained to review data provided by EFB just as critically as they review data provided on paper. Crews should understand the source of the data and not place excessive faith in it just because it is generated electronically. Pilots should be especially careful to ensure that the data are applicable to current operating conditions.

Examples

An EFB might contain a database of information related to the particular aircraft it is on for the purpose of calculating flight performance. This database may be updated infrequently, but it should be checked carefully for completeness and accuracy before it is installed on the EFB.

While parked at the gate, a maintenance person may wish to use the EFB while the crew is out. When the crew returns they may do one of the following sample procedures for reviewing EFB flight performance calculations that were completed prior to leaving the unit unattended:

(1) Clear all entered data and re-enter it themselves.

(2) Review and check all data entered from a remote source, such as Flight Dispatch, then re-enter all locally entered data. (To aid the pilot, all entered data could be highlighted on the EFB.)

(3) If a calculation is based on the locally entered data, the pilot could review all the calculation steps, even if no data entry was required; i.e., he/she could review all the raw data on which the calculations are based.

Evaluation Questions
- What are the procedures for ensuring that the EFB databases are accurate, up-to-date, and uncorrupted?
- What are the procedures for reviewing entered EFB data?

2.4.16 Updating EFB Data

Training/Procedures Recommendations

> - Procedures should be developed for handling EFB data updates (e.g., for software or databases) to ensure that the applications and data are not corrupted in any way during the updating process. These procedures should be consistent with existing standard operating procedures.
>
> NOTE: AC 120-76A contains several provisions for updating databases and software on an EFB. This AC applies to all operations other than Part 91 (excluding subpart F, to which the AC does apply). The intention of the requirement here is to make operators and users who are not subject to AC 120-76A aware of the importance of procedures for updating EFB data and software. For all others, the guidance in AC 120-76A supersedes this requirement.
>
> NOTE: In air transport operations, EFB updates will most likely be handled by staff other than the flight crew. In other operations, flight crew may be more involved in the EFB update process.

Equipment Suggestions

> - There should be a plan for how often and when updates will be made.
> - Manufacturers should have a plan for how to handle modifications to the EFB after its initial purchase and installation. This plan should be understood by the customer.

Problem Statement

When an EFB arrives direct from the manufacturer, it will probably have undergone checks to ensure that the applications and data on it operate correctly. Once the operator is responsible for updating the EFB on a regular basis, however, it is important that standard practices are employed to ensure that the EFB continues to operate correctly. Flight crews and/or maintenance personnel who update the software need to use standard practices so that they do not introduce any errors or corrupt the database in other ways that would cause the EFB to fail, or degrade its performance.

It is important that the updates to EFB software/databases are made correctly and in a timely fashion. If the process breaks down, the EFB may not be usable for a given flight, or, if the information is essential, a flight delay may be incurred while the EFB equipment is fixed.

Customers expect that EFB software and databases will be upgraded and customized multiple times over its lifetime. It is important that there is clear communication between the manufacturer and the customer about how upgrades and customizations are to be performed. The responsibility for performing the EFB modifications should be clearly assigned.

Examples

As new versions of flight crew manuals are released, they should be loaded into the EFB. The manuals could be loaded via maintenance procedures, crew procedures, or even via automatic data link.

When a new version of an EFB software application is installed, the procedure for installation could include a final check that all files related to outdated versions of the applications are removed. A final check on the installation of the upgrade could include running a test of the new software on the EFB.

Updates could be made on the same schedule as updates to the FMS (every 28 days) in Part 121 operations, so as to limit aircraft down time.

Evaluation Questions

- What is the process for updating EFB software/databases? Who will do the updates, and when?
- How will the updates be documented for the crew who has to check approval for use in flight?

General EFB System 59

2.4.17 Crew Confirmation of EFB Software/Database Approval

Equipment Requirement

> - The EFB should provide the latest revision information to the crew upon request.

Equipment and Training/Procedures Recommendation

> - A procedure should be in place for crews to either confirm the revision numbers and/or dates of EFB databases and software installed on their units for each flight, or for crews to be notified that an EFB self-test has identified data sources that are out of date. (AC 120-76A, Section 10 f (2))

Training/Procedures Recommendations

> - Procedures should specify what action to take if the applications or databases loaded on the EFB are out of date. (AC 120-76A, Section 10.f (2))
> - Flight crews should not be required to confirm the revision dates for databases that do not adversely affect flight operations, such as maintenance log forms, a list of airport codes, or the Captain's Atlas. (AC 120-76A, Section 10.f (2))
> - The procedure for checking whether the EFB software/data is approved should be consistent across the different applications available on the EFB, and consistent with airline/operator standard operating procedures.
> - Change information explaining the updates made in the latest revision of the EFB software and databases should be provided to the crews.

Problem Statement

Whether maintenance crews or flight crews load databases and software onto the EFB, the flight crew is ultimately responsible for ensuring that they use approved information to conduct the flight. Approved information is usually the most current, although there may be special cases where older information is still approved for use in flight. A procedure needs to be in place to make sure that this task is as simple and error-proof as possible for the crew.

Since EFB software and databases may change at different times for the different applications, making sure that approved information is in place can become even more complex if the procedure for checking approval differs between applications. Therefore, it is best to keep the procedure consistent across the different EFB functions. Ideally, approval information for all EFB software/databases would be available in one place.

Examples

One way of checking whether the EFB contains approved information is to provide crews with valid dates for the EFB. Crews could check that the current date is valid before each flight.

An EFB could automatically check the date and provide the pilot with a message stating whether the database was approved or not for the current flight. The automated checking routines would have to be verified themselves. For example, the EFB could check the current date and time against the EFB valid dates, but in this case, the system's knowledge of current date and time should be validated (e.g., against an independent GPS time stamp).

Evaluation Questions

- What is the procedure for ensuring that data in use is approved for use in flight? Is the procedure for checking the EFB data approval consistent with standard operating procedures?
- Can the crew request revision information from the EFB? Is the revision information presented clearly?
- Are procedures in place so pilots know what to do if the database is not approved for use in flight?

2.4.18 Links to Related Material

Equipment Recommendations

> - If related information is accessible, a consistent philosophy should be used for determining how different types of information are accessed. Similar types of information should be accessed in the same way.
> - If related information is accessible, it should be easy for the user to keep track of how to move between the different topics. In particular, it should be easy for the user to return to the place that he/she started from easily.

Equipment Suggestion

> - Access to related information (e.g., more detailed information, or definitions of acronyms and terms) could be provided through links to electronic documents, pop-up information, or other similar techniques.

Problem Statement

While pilots are very familiar with the paper documents they use (such as charts, documents, and checklists), it is desirable to increase the usability of these documents by building in links to related information. If links are available, they may allow the pilot to obtain all the necessary information from one source, rather than requiring him/her to open multiple documents or document segments to obtain the information. Links are also helpful when the pilot does not have a focused question in mind, but rather is trying to obtain more general information.

The type of information being related should determine the access method used. A consistent access philosophy will help users to anticipate what they will find if they access linked information.

Examples

Examples of access methods include hyperlinks to other sections of a document, pop-up panels that provide *brief* definitions of terms and acronyms (e.g., "balloon help"), context-sensitive help, and navigation buttons or tabs displayed on the screen.

Definitions of words and acronyms can be provided through pop-up panels. The pop-up panels could be activated when a word or acronym is selected, or they could be identified by a special icon and then accessed when the icon is selected.

Hyperlinks are useful to enable the user to access additional information on a referenced topic. They are especially appropriate for supporting navigation from one location in a document to another document section that provides more detail.

Navigation buttons or tabs are useful for supporting access to complementary types of information. For example, the operations manual, which dedicates a chapter to each aircraft system, could display navigation tabs at the bottom of the screen that support access to checklists and MEL information from each system chapter.

If a consistent philosophy of access to information is implemented, all definitions could be accessed in one way (e.g., pop-up panels), and all links to other document segments could be implemented in a different way (e.g., navigation buttons). The user then expects different types of information when he/she clicks on a term as opposed to when they click on a navigation button.

It is possible that different visual information will be relevant and useful at different display scales of electronic charts. Examples include terrain rendering, holding quadrants, and airport runways. One approach to providing flexible levels of detail would be to automatically render the information that's relevant to a particular scale but let the pilot ask for more detail of an object, perhaps by hovering over that object or selecting the object and requesting more detail from a context-sensitive menu.

Charts, or portions of charts could also be linked. For example, charts for parallel runways could be linked so that it is easy to pull up the other runway if a sidestep maneuver is requested late in the approach.

Evaluation Questions
- Is access to related information supported?
- Are similar types of information accessed in the same way?
- How complex is it to return to the place where the user started from?

2.4.19 User-Interface Customization

Equipment Requirement

> NOTE: Examples of user-interface customization include changing color conventions, font conventions, icon styles, and input device parameters. Adjustments to the display to improve viewing of the content (e.g., by adjusting brightness, changing font size, zooming, or panning) are not considered to be customization of the user interface.
>
> - If the application supports user-interface customization by the end user, it should also provide the end user with an easy means by which to reset all customized parameters back to their default values.

Equipment Recommendations

> - For Part 121 and 135 operations, default settings for the user interface should be customizable only by an appropriately authorized administrator.
> - For Part 91 operations, default settings for the user interface should be specified by the manufacturer. In addition, the user should be allowed to configure their own default settings as well.

Equipment and Training/Procedures Recommendations

> - In general, user-interface customizations that affect user interface conventions for the system (e.g., color settings or fonts in use) should be limited to pre-flight (i.e., prior to engine start).
> - In some EFBs, color conventions will change based on day/night mode, and this change may occur mid-flight. In this case:
>
>   (a) The shift from one convention to the other should be obvious to the crew
>
>   (b) The crew should be familiar with both day and night conventions.
>
>   (c) Differences between the day and night color conventions should be analyzed to ensure that the one set of color definitions are not in direct conflict with the other.

Equipment Design Tradeoff(s)

> The extent to which the information display can be customized should be carefully evaluated to ensure that degradations in usability, legibility and readability do not occur. It may be appropriate to limit the types and range of customization for use on the flight deck relative to that which is typically provided for a standard graphical user interface.

Problem Statement

It can be helpful for the operator of an EFB to have some flexibility in the definition of the user interface conventions, rather than to have to use manufacturer provided default settings. Some examples of conventions that operators may want to adjust include color conventions, font style conventions, icon styles, and parameters for input devices. Note that these parameters are at a higher level that what the end user will typically need to modify in use, such as display brightness. In some cases tailoring by the operator can improve upon the equipment defaults, but in other cases, it could result in degraded information appearance. Therefore, it may be necessary to consider carefully what user interface parameters may be manipulated by the operator and the range of manipulation that should be supported.

Customization of the user interface will likely be handled differently by airline operators versus private operators. In the airline environment, customization of the EFB user interface should be addressed similar to

General EFB System 63

other EFB configuration issues (e.g., verification of EFB data, 2.4.15). In private operations, the end user may have a more individual level of responsibility for customizing the user interface.

To ensure that basic usability of the device is maintained, there should be a way for the end user to return to a set of default user interface settings. For airline operators, the default could be configured by an airline staff person. For a private operation, the user may want to return to the manufacturer-specified defaults, or their own custom default settings.

User-interface customization should be limited to pre-flight because it may be confusing to have the conventions change during the flight and because making any changes may take significant crew attention. Note the exception for day/night color conventions discussed above.

Examples

Personal computers allow users to manipulate resolution, background and typeface colors, font and document size, and a host of other parameters. The flight deck environment may not be the place to experiment with this broad flexibility. Each customizable parameter should be assessed as to its potential for reducing the readability and legibility of the information presented.

Unlike most office computers, EFBs may often be used by more than one user. Customizable parameters could be stored centrally, thus reducing the amount of time spent manipulating the parameters in the first place and allowing for a single reset button that would return all parameters to their default settings.

Evaluation Questions

- Does the EFB provide an easy means for resetting all parameters to their default values?
- Can the manipulation of a display parameter produce a significant decrement in the appearance of the displayed information?

2.5 Hardware Considerations

2.5.1 Pointing and Cursor Control Devices

Equipment Recommendations

> - Input devices, such as keyboards or cursor-control devices, should be selected and customized based on the type and complexity of the entries to be made and flight deck environmental factors that could affect the usability of that input device, such as turbulence and even normal vibration.
> - The performance parameters of input devices should be tailored for the intended application as well as for the flight deck environment.
> - There should be a way to rest and/or stabilize the hand when actively using the pointing or cursor control device.
> - Active areas on the display (e.g., touch screen controls) should be sized to permit accurate selection with the pointing/cursor-control device in the flight deck environment under all operating conditions (e.g., turbulence).

Problem Statement

Pointing and cursor control devices are used to identify and select a specific point on the screen. Some pointing devices (e.g., a track ball, displacement joystick, touch pad, and mouse) operate smoothly; they allow the user to make fine selections quickly, easily, and with a great deal of flexibility in the path they take from one place to another on the screen. Other devices, such as joy pads, which are common on personal digital assistants, are commonly used for discrete movements. They allow the user to either move up/down, or right/ left, and act like multi-directional toggle switches. Force sticks (sometimes called "top hats") are another type of pointing device. These devices move the cursor in response to the force the user exerts, right, left, up or down. They are typically small controls that are integrated onto a keyboard.

The user will interact with EFB functions via some input device, such as a track ball, touch pad, rotary knob, keyboard, or soft keys. The input device should be matched to the type and complexity of the entries to be made; the entries to be made will vary by the function that is being performed. If the input device is poorly matched to the task, not only are errors more likely, but the task with will take considerably more time to complete, and users will become frustrated, and potentially even distracted from higher priority tasks. In addition, it is important to maximize the ease of using any particular input mechanism. For example, some input devices will benefit from having a place to rest or stabilize the hand. Also, touch screen controls need to be sized for accurate selection under all operating conditions.

Examples

One simple type of input is to select a text hyperlink on an electronic document using a cursor control. More complex input, such as numeric data for performance calculations may require the use of a simple, or full-featured, keyboard. If the EFB does not support entry of free text, a full keyboard may not be necessary. Control of continuous functions, such as screen brightness, is best achieved by means of a continuous control, such as a thumbwheel or knob.

As an example of optimizing an input device, consider the case of a touch screen; the active areas may need to be larger in a flight deck environment than they would be in a stable environment to promote accurate data entry in turbulence. This will in turn impact the size of the display, and the size of the unit itself.

The Boeing Company published a paper on their efforts to design input devices for the 777 that contains good information on design of transport category aircraft input devices. (Crane, Bang, and Hartel, 1994; see References.) One of their recommendations is that the dimensions of active areas on the screen be a minimum of 3/8-inch in height and width, but ½-inch on each side is preferable for operations in turbulence.

Evaluation Questions
- Can crews use the input mechanism accurately and reliably for the least common types of data entry without an unusual level of skill, patience, or practice?
- Can the user position the pointer/cursor mechanism (if any) quickly, reliably, and repeatedly under all flight conditions (e.g., turbulence, darkness)?

2.5.2 Hardware Controls

Equipment Requirements

- All controls should be properly labeled for their intended function. (14 CFR 23.1555, 25.1555, and 27.1555)
- All soft function keys that have an associated action should be labeled for their current intended function.

  NOTE: Soft function keys are physical buttons whose actions can be reassigned via software.

- Soft function keys that are inactive should either not be labeled, or use some kind of display convention to indicate that the function is not available.

Equipment Recommendations

- Physical function keys should provide tactile feedback to the user when pushed.
- If the function key records lengthy activations as separate events (e.g., key repeats), the software should filter out these events if they occur too closely together for the user to have intentionally entered them as separate actions.
- Soft function key labels should be drawn in a reserved space outside of the main content area.
- Soft function keys could be used to select one of several available functions. When the same type of function is accessed from different points in the software (e.g., exit/quit the application), the common function should appear on the same physical function key whenever possible (e.g., top right).
- Labels should be used consistently throughout the software.
- Labels used to identify the action associated with a soft function key should be clear to the user and brief.
- Standard abbreviations may be used in the labels, but the user should not have to learn new or ambiguous abbreviations.
- Labels should be located near the controls they identify. They should not be confusingly close to other labels, or to other controls for which they do not apply.
- Labels should be drawn in horizontal, not vertical text.
- Physical controls should generally be co-located with the display.

  NOTE: Some general-purpose inputs devices, such as keyboards, may need to be separate from the display unit.

- Controls should be organized so that the most frequently used controls are placed at the most accessible locations.
- Inadvertent activation of controls should be deterred.

Equipment Suggestions

> - If there are many controls in a small space, they may need to be grouped according to function, and/or arranged in order of use.
> - The EFB display and control hardware should meet the requirements given in the FAA Human Factors Design Standards for Acquisition of Commercial-off-the-shelf Subsystems, Non-Developmental Items, and Developmental Systems—Final Report and Guide (Chapter 6, Controls and Visual Indicators). This reference covers topics such as dimensions of the control surface, spacing between controls, forces required to activate controls, and appropriate displacements for displacement controls such as toggle switches.

Equipment Design Tradeoff(s)

> If the display glass is set inside a bezel, the depth of the bezel frame can introduce a perceived misalignment between soft key labels and physical function keys when viewing the display off-angle (i.e., parallax errors).

Problem Statement

Some examples of physical controls are push buttons, rotary knobs, thumbwheels and toggle switches. Soft function keys are physical push buttons whose functions can be reassigned to a variety of actions via software. A common example of soft function keys in the flight deck are the line select keys on the FMS CDU.

Expectations about the control's action play an important role in ease of use. For this reason, physical controls need to be labeled so that the user knows what to expect when using them. The labels should be clear, but brief. Labels should also aid the flight crew in identifying, interpreting, and following procedures. They should describe the control functions and provide a memory cue to the user that facilitates interaction. Soft key labels should be drawn in a reserved area of the screen so that they neither obscure the application information on the EFB, nor are themselves obscured.

Many other parameters of the controls should also be designed carefully, such as dimensions of the control surface, spacing between controls, forces required to activate controls, and displacements for displacement controls such as toggle switches. If these factors are not taken into account, the physical controls could be either too difficult, or too easy to activate.

Poor design of hardware controls can increase pilot workload by requiring the pilot to check his or her entries more carefully, and by having to correct incorrect entries more often. Tactile feedback can help ensure that the pilot is aware of his/her entries.

Examples

When soft keys are consistently mapped with a particular action, users can associate that key's location with that action, making the software easier to use. If a function is mapped to different keys at different times or in different states, errors may occur because the user expects the function to be assigned to the same key all the time.

Multiple entries are sometimes registered by the hardware when a user holds down a button longer than usual, or when the user's finger hits the button twice inadvertently (e.g., in turbulence). A good rule of thumb is to discard multiple entries that occur within 300 milliseconds of each other, which is a typical length for the time between fast, discrete, intentional movements.

Some suggestions to reduce inadvertent activation of controls include (a) placing fences between closely spaced controls, (b) making the upper surface of keys concave to prevent slippage when selecting that key, or (c) the size of the control could be increased to allow for accurate selection.

It may be a good idea to dedicate a function key for functions that require quick access (e.g., return to the top level menu). Not only should that function always be on the same function key, but that function should not be used for any other purpose.

In order to keep labels clear and brief, the designer should consider whether any information is repeated often. For example, if all the labels are for access to different charts for a single airport, the airport identifier should only appear once on the page, as an overall reminder. The airport identifier should not be repeated in the label for every function key on the screen.

Hardware controls (e.g., thumbwheels, knobs, toggle switches) should be selected appropriately for the task. For example, it is more likely that a thumbwheel will be inadvertently adjusted than a rotary knob with detents, so thumbwheels should not be used when accidental manipulation is to be strongly discouraged.

Evaluation Questions

- Are controls labeled consistently and briefly for their intended function?
- Can the user easily perform the most common types of input in any operational environment?
- Do physical keys provide tactile feedback?
- Are inadvertent multiple entries discarded?
- Is inadvertent activation of controls deterred?

General EFB System   69

2.5.3 Display

Equipment Recommendations

> • The physical nature of the display screen should minimize the likelihood of becoming obscured (e.g., by repelling smoke or dust particles, and by being easy to clean).

Equipment Design Tradeoff(s)

> Some basic features of displays that affect their usability include:
>
> a) Resolution
>
> b) Physical dimensions of viewing area
>
> c) Number of colors supported
>
> d) Range of brightness (lowest value and highest value)
>
> e) Off-axis readability (at what point do colors and other display features wash out?)
>
> The technology used to generate the display (e.g., active matrix LCD, or passive matrix LCD, and the type of backlighting in use) can affect the visual quality perceived by user in different ways. In some cases, there are techniques (such as anti-aliasing, described below) that can improve the perceived visual quality of the display.
>
> Liquid crystal displays (LCDs) are particularly susceptible to poor off-axis readability. As the off-axis viewing angle is increased, different colors will wash out and become impossible to discriminate. Displays of video information, which are usually of relatively constant brightness, may be particularly susceptible to washing out earlier than other displays with brighter, more contrasting colors.
>
> Most conventional LCDs are strictly backlit, and their visibility in sunlight and glare depends on how bright the backlighting can be set. However, there is a reflective LCD technology that does not rely on backlighting for visibility in high ambient light. This display can be viewed in sunlight with the backlighting off, and can be backlit for viewing in low visibility conditions.
>
> Touch screens are particularly subject to display degradation during normal use because finger oils accumulate on the display. The finger oils can result in distracting smudges that produce glare and strange colors caused by refraction of the backlight through them. The finger oils can also attract dust and dirt to the screen.
>
> Consideration should be given to the long-term display degradation as a result of abrasion and aging.
>
> The display glass is often inset in a bezel. The depth of the bezel frame can impact the user's ability to see the whole drawing area on the screen. (The edges of the drawing area may not be visible when viewing the display off-angle.)

Problem Statement

The technology used to create an electronic visual display can significantly affect its usability. Different display technologies are susceptible to different types of usability problems. Each display should be evaluated along many dimensions for its overall suitability for the intended functionality of the EFB.

Examples

Each pixel is controlled independently on an active matrix LCDs, but not on a passive matrix LCD. As a result, images on a passive matrix display are typically a little fuzzier and less well defined than those on an active matrix display because of how the images are generated. Also, a high contrast transition on a passive LCD can cause an artifact along the rest of the display. For example, a black box on a white background may produce gray lines that run across the display in both directions from the box because the voltage difference required to render the black box may leak across all the pixels along those rows and columns. Display flicker can also be more noticeable with passive matrix LCDs.

Anti-aliasing smoothes out diagonal lines on the display by adding pixels with intermediate color and brightness values between the figure or line and the background to fill in the gaps where the edge would appear jagged. The more colors available to the display, the more fine-tuned the anti-aliasing can be, and the smoother the resulting rendering will appear.

Off-axis readability is often important for effective crew coordination. It is common for, say, a first officer to enter data into a system such as the FMS CDU, and for the captain to try to follow what the first officer is doing. Any display that is not exactly replicated for both seats should be easily viewed from either seat without requiring major adjustments to body position. Often, the pilot who is trying to watch what the other pilot is doing is responsible for continuing to fly the airplane, so leaning over the pedestal and other major posture changes are undesirable.

A protective cover may be desirable for EFBs that are portable, in order to minimize wear on the display screen.

Recommendations for cleaning the display screen could be provided to customers by the manufacturer.

Evaluation Questions

- Does the display provide enough clarity for use of the intended applications? Consider its resolution, brightness, off-axis readability, etc.
- What kinds of artifacts can appear on the display (e.g., ghost images or lines, jagged lines, or fuzzy images)?
- Does the display continue to be usable after prolonged use in the flight deck environment?

General EFB System

2.5.4 Accessibility of Hardware Components

Equipment and Installation Recommendations

- Hardware components that are designed for routine use by the crew (e.g., cables, cable connectors, ports, disk drives, data card readers) should be easily accessible and safeguarded to prevent damage during normal use.
- Controls that are not intended for use by the flight crew should not be easily accessible.
- When in use, hardware components should be physically oriented so that the user can use them without visual contact.
- Locking connectors (e.g., those that use screws or clips to secure the cable) should be usable without special tools.
- Cables that connect in a particular orientation should be designed to prevent insertion in any other orientation. If there are multiple cables, each cable should fit only into its intended slot; it should not be possible to insert a connector into the wrong slot.
- Any cable should not hang loosely in a way that compromises task performance and safety.
- Flight crew members should easily be able to secure any cables out of the way during aircraft operations (e.g., cable tether straps).
- If multiple cables are present, they should be color coded for ease of visual identification.

Problem Statement

Any hardware components that are expected to be used by the crew (e.g., cables, cable connectors, ports, disk drives, data card readers) need to be robust enough for repeated, occasionally rough, use by non-experts. If the components are not robust, the hardware may fail quicker than expected under real-world operating conditions. If the components are not well designed from the user's perspective, they will undergo rougher treatment by users. Also, in selecting placement of hardware components, designers should consider how the crew will access them and whether there will be any impact on flight task performance and safety.

Examples

Connectors may be present on the EFB for access to power, or data from the aircraft. If the pilot routinely needs to engage and disengage these connectors, they need to be designed for ease of use by a non-expert hardware user.

Locking connectors are likely to be preferred for use on EFBs because they will be resistant to falling off in the aircraft environment. However, the crew may need to be able to engage and disengage the connector, and should not be required to use tools for this task.

Thumbscrews are preferred over screws that require use of a screwdriver if they need to be used by the flight crew.

Evaluation Questions

- What are the hardware components that are routinely used by the crew? Are they easy to access? Is there any impact on flight task performance or safety?
- Are the connectors easy to use? Consider how long it takes to make the connections, how likely errors will be, and whether any special tools are required.
- Are the hardware components robust enough for use in the flight environment? For example, will connectors stay in place after lengthy use in a vibrating environment?

2.5.5 Keyboards

Equipment Recommendations

> - Keyboards should be chosen and designed based on the specific pilot tasks they support. QWERTY keyboards should be used for tasks that involve free text entry. Numeric keypads are best suited for tasks involving significant numeric entries.
> - Keyboards should provide sufficient tactile feedback of key depression to ensure that the user knows when the key has been actuated.
> - There should be a position in front of the keyboard on which to stabilize and rest the hand for reliable use in turbulence.

Problem Statement

The type of keyboard used for data entry should be appropriate for the pilots' tasks. Some examples of keyboard options include: QWERTY (full), or alphabetic; wireless or wired; full size or reduced size; keypads include numeric keypads, or arrow keys only; and finally, thumb keyboards (e.g., as seen on personal digital assistants).

For tasks involving free text entry, a QWERTY-style keyboard is preferred. Any departure from that, such as an alphabetic layout, or a thumb-operated keyboard, involves compromises. However, if text data entry is limited to a small number of discrete entries, the compromise in usability (e.g., slower typing speed, more errors in entering text) may be justified by the increased portability of the device.

For any task involving numeric data entry, a dedicated numeric keypad is best. Any departure from that, such as an embedded numeric keypad (where the numbers are secondary functions of the alphabetic keys) involves compromises as well, again in the form of slower speeds and lower accuracy in entering data.

Any keyboard used in flight should have an adequate place to rest the palm(s) for stable operation in turbulence. The action of the keys should be firm enough to reduce the likelihood of unintended actuation, and the keys should have a positive "click" when the key actuates so there is no doubt about whether the input was made successfully or not, even in turbulence.

Examples

Extensive text entry requires a full QWERTY keyboard positioned for use with both hands. Less extensive data entry can be accommodated with either a QWERTY keyboard or an alphabetic keyboard. If the keyboard is not intended to be used with two hands, an alphabetic layout is often sufficient, but users may still prefer the QWERTY layout because of familiarity.

Some sub-notebook computers have reduced size QWERTY keyboards. These are acceptable to some users, but not to others.

Some PDAs and pagers have thumb-operable keyboards; these are small, usually round keys arranged in a QWERTY configuration but rounded around the center front of the device so that when the thumbs are in the center front of the case, the keys are arrayed around them.

Numeric keypads are sometimes embedded in QWERTY keyboards and accessed by changing the keyboard function or mode. However, since keys arranged in the QWERTY configuration are usually offset from each other row by row, and numeric keypad keys are usually arranged in straight rows and columns, embedding the numeric keys in the offset alphabetic keys can make it very difficult to use the numeric keys without looking, since the spatial relationships between the keys are different from the dedicated keypad configuration.

Evaluation Questions

- Is the keyboard appropriate for the task?
- Do the keys provide positive tactile feedback that can be felt even in turbulence?
- Is key action firm enough to resist unintended actuation?

3 Electronic Documents

The first function proposed for EFBs was to support electronic versions of flight-related documents, such as reference manuals and other operating documents. The goal is to convert paper documents into electronic documents while retaining, if not enhancing, readability and access to data. For example, electronic documents could be cross-linked to make it easy to read about a single topic that is addressed in more than one document. Certain paper-based attributes such as tables of content, indices, and cross-references can be significantly enhanced by automatic linking in an electronic media (i.e., hypertext links). In addition, electronic media can allow full text searches that enable the user to access information across multiple documents. Electronic documents are also expected to be easier to update and distribute.

Both paper-based and electronic document systems require information to be logically organized and structured to ensure the user can access and use the information with minimal effort. Poorly organized and structured paper-based document systems will not be enhanced by simple conversion to an electronic media; such information should be reorganized/structured before conversion.

In Section 3.1 the type of documents that are being considered here are clarified and the many options that electronic documents might or might not support are reviewed. Considerations for electronic documents begin in Section 3.2, with General topics. Layout and Appearance issues are presented in Section 3.3. Navigation and Search issues are presented in Section 3.4. Finally, optional (more advanced) electronic document features are presented in Section 3.5.

3.1 Background

3.1.1 Type of Documents Addressed

In the United States, pilots carry at least three types of documents in their flight bag: manuals, checklists, and navigation publications. Only manuals are covered in this chapter. Checklists are covered in Chapter 4. Approach charts, just one type of navigation publication, are addressed in Chapter 5.1.8. The manuals considered in this chapter include, for example:

- Pilot's Operating Handbook (POH), which contains information on the aircraft and its systems
- Flight Operations Manual (FOM), which contains airline policies and procedures, including emergency procedures
- Airport Analysis and Aircraft Performance Manual, which contains information about specific airports and runways
- Minimum Equipment List (MEL) and configuration deviation lists, which contain information about operational restrictions stemming from limitations in case of partial or full failure of various aircraft instrumentation and systems

The FAA mandates that the documents listed above be on board the aircraft. The FAA also approves or accepts their content, and reviews any changes. Other similar documents, such as the Aeronautical Information Manual, the Federal Aviation Regulations, and aircraft maintenance manuals, could be also be made available on an EFB. (AC 120-76A contains a more detailed list of flight-related documents in its Appendix A.)

These flight manuals are primarily reference documents that are used relatively infrequently to find specific information, although they are sometimes used for studying topics in depth. The manuals are updated relatively infrequently (i.e., over a period of days, not hours). They are not used interactively in the way that electronic charts or electronic checklists will be used, where the pilot routinely customizes the display, or enters data. Also, these manuals are not usually accessed in time-critical situations.

3.1.2 Features of Electronic Documents

Electronic documents may vary greatly from one another. They may vary in how the document is displayed, whether the view is customizable, how the user enters information and commands, how the user enters text (if at all), and what level of support there is for multimedia. All of these features are affected by the framework used to create electronic documents (i.e., the library of software routines). The framework is usually based on the operating system of the EFB device, though it may not be.

Advanced and basic operating systems are compared in Table 3-1 below. In general, advanced operating systems support advanced document features, but this is not always the case. For example, if the operating system supports windows, the user usually is given the option to reposition and resize the windows. However, it is best to examine each implementation of electronic documents feature by feature.

Advanced and basic electronic document features are compared in Table 3-2 on the next page. Display and input hardware limitations that may affect the user interface of electronic documents are also listed in Table 3-2. Not all systems fit directly into either the "advanced" or "basic" categories. In some cases, it may be possible to support both advanced and basic functionalities (e.g., a screen may support both cursor-based input and function-key input). The categories merely illustrate the range of possibilities.

Another independent factor that affects electronic document functionality was already mentioned in Chapter 1: Is the EFB integrated with other flight deck systems? If it is, then electronic documents that are related to a given system condition (e.g., failure of an item) could be suggested to the crew from built-in logic. In this case, the electronic document functionality is a decision support tool, not just a data-access tool.

Some examples illustrate the spectrum of electronic documents. First, consider a simple display that shows unformatted text and static graphics in a single window or frame. This implementation might marginally serve as an information repository, but it would probably not be easier to use or more functional than the originating paper document, and could be worse. Without information being logically and visually "chunked," neither paper nor electronic system would probably satisfy pilots' needs.

A second example would be to convert an existing, well-crafted paper document to an electronic page-based display system (e.g., Adobe Acrobat). This would preserve the original document's organization and structure and additionally provide benefits of electronic media (e.g., hypertext linking, full-text search). Some drawbacks include certain topographical considerations such as typeface optimization for screen presentation, and screen size display capability.

The most advanced electronic document technology to date is based on markup languages, such as XML. Markup languages give the document automatic functionality such as word wrapping and/or zooming without loss of quality and placement of graphics. XML documents are also more flexible in terms of searches as compared with PDF. However, XML does not preserve the paper document's page structure. Sections are preserved as opposed to pages.

Operating System	Advanced: Industry standard (e.g., Windows, Macintosh, or Unix) or equivalent proprietary system. The operating system supports a graphical user interface with windows and menus. There are standardized dialog boxes with standard interface elements, such as text entry boxes and command buttons. There are standardized dialog boxes for displaying system status and alerts, which are compatible with other types of dialogs boxes. The operating system is typically designed for use on personal/business computers outside of the aviation environment. It can run more than one software application at a time. Text formatting (e.g., changing font color, size, or underlining) is fully supported.
	Basic: Compatible with industry standard operating system to some degree. The basic operating system could be more or less compatible with industry standard. For example, the graphical user interface could be limited, or unsupported. Basic dialog boxes may be supported, such as modal system alerts or simple user entry dialogs (e.g., with yes/no responses), but complex user entry dialog boxes are not supported. Typically, these systems have multi-function keys, which are physical buttons whose functions are configurable via software. Text formatting could be limited or unavailable.

Table 3-1. Differences between advanced and basic EFB operating systems.

Display Area	**Advanced**: Documents are displayed in windows that can be repositioned on the screen and resized. The windows support scrolling text and color. Page layout can be formatted similar to a paper page on the window. More than one window/document can be open at a time and overlapped, with one window covering part of another window. May allow the user to customize the display in some ways. For example, windows could be resized or repositioned, or font properties could be modified if allowed. If markup languages are used, the text will automatically be reformatted for the current visible display area.
	Basic: Either (a) or (b). Neither typically allows the user to customize the display.
	(a) Regions/Frames. The screen area has selectable regions/frames. The regions may support scrolling, but cannot be repositioned or resized individually. The basic display area does not support overlapped display regions.
	(b) No Regions/Frames. The screen area does not have selectable regions/frames. Scrolling, if supported, affects entire screen contents.
General User Input	**Advanced**: Users can point to any area of the screen and activate that point (e.g., by using a mouse, touch pad, or pen input). If that area is active, a change will occur; for example, a menu may drop down or pop-up, or a cursor may be moved to that position.
	Users may select actions via graphical (pop-up or pull-down) menus. The list of menu options is out of view until called up by user.
	The on-screen cursor may allow the user to select regions of text, for example, for printing or copying. The cursor could also be used to activate hyperlinks or to activate pop-up areas for definitions, help, or related information on a topic.
	There may be standardized dialog boxes with standard interface elements, such as text entry boxes and command buttons. There are standard practices for designing dialogs for confirmation of important actions and collecting user-entered data. Dialog boxes could be modal (i.e., no other user actions are accepted until dismissed) or modeless (i.e., other user actions are processed normally while the box is visible).
	Basic: Function keys are used to select menu items, activate commands, and navigate between menus and screens. The same buttons could move the cursor or scroll text. The user selects from a visible list of alternative actions. (There might not be any pop-up or pull-down menus.)
Text Entry	**Advanced**: Users can enter free text easily (e.g., through a keyboard, or even hand-writing recognition software). The free text capability could be used, for example, to enter annotations, or to enter key words for search.
	Basic: Either (a) or (b)
	(a) Keypad-style text entry. Text entry is possible but usually cumbersome because more than one key press could be needed to enter a specific letter. For example, letters could be entered with 10 push buttons, configured as on a telephone. (To enter a "B", press the "1" key, which corresponds to "A", "B", and "C", twice.)
	(b) No text entry supported.
Multimedia Support	**Advanced**: Includes support for audio and visual animation, such as video clips, and/or sound recordings.
	Basic: Limited or no support for audio and visual animation. For example, the audio may be limited to a beep, or a few pre-programmed sounds.

Table 3-2. Comparison of advanced and basic electronic document features.

3.2 General

3.2.1 Consistency of Information Structure

Equipment Requirement

> - The information structure of an electronic document (e.g., its outline, section headings, and table of contents) should be consistent with the hard copy version of that material, if a hard copy exists.
>
> NOTE: Visual structure of the document (e.g., page layout and formatting) is discussed in 3.3.1.

Equipment Design Tradeoff(s)

> When converting paper manuals into electronic manuals, it may be possible to streamline the electronic document by deleting content (e.g., a section) that is not relevant to the aircraft in which the EFB will be used. The document could be customized either when the data are loaded onto the EFB, or through entry of specific aircraft information into the EFB. Integrated EFBs, which are connected to other aircraft systems, could potentially acquire aircraft-specific data directly through a data bus.
>
> When content is considered for deletion, designers should ensure that that data are not necessary in any unusual circumstances. Also, if entire sections of text are deleted, they should still be called out by a section heading, so that consistency with the more complete paper document is maintained.

Problem Statement

Pilots are very familiar with the information structure of hard copy versions of required documents. They use this background knowledge when searching for information through a table of contents or index. If the information structure of electronic documents differs from that of the paper documents, pilots may become confused and then have trouble locating material in either or both types of documents. (Note that "information structure" refers to the logical structure of the information in the document, as opposed to "visual structure," which refers to the visual appearance of the document.)

In cases where the operator determines that electronic documents are better designed for pilot use than the paper versions, it is appropriate to update the content of the paper manuals. Consistency between the two documents' information structures is required, but evolution towards a better structure for all versions is encouraged.

Examples

The section headings and section numbers for any reference manual should be the same in both the electronic and paper versions of the document. It may not be necessary to keep page numbers on the electronic version, as long as the content can be referenced in other ways, such as through section headings.

An example where customization of electronic documents could be useful is when there are multiple aircraft models or series that are very similar. For example, there is one paper manual for both Boeing 757 and 767 aircraft. Within this document, there are notes next to any material that applies to only one of the two types. An electronic version of this document could be configured to display only the material relevant to the aircraft model of interest.

Evaluation Questions

- Is the information structure of the electronic document consistent with any hard copy version of that material?

3.2.2 Training Needs

Training/Procedures Recommendation

> - If there are any differences in the content or information structure of the paper and electronic documents, users should be made aware of these differences.

Training/Procedures Suggestion

> - Basic procedures for moving through an electronic document should be demonstrated during initial training, even if they are covered only briefly just to ensure that even non-computer-savvy users are brought up to speed. These include how to choose a document for display, displaying a selected section, and moving between pages in the same document segment. If multiple documents can be open at the same time, training should also address how to move between documents.

Training/Procedures Design Tradeoff

> Users are likely to need more training to work with electronic documents that support advanced features.

Problem Statement

To support compatibility between the paper and electronic versions of the same documents, electronic documents should implement the information structure of the paper document (see 3.2.1). However, users should be aware if there are, in fact, any content differences between electronic documents and paper documents. Users should also understand how to access and display different segments of the electronic document(s) available on the system so that they do not have trouble getting to a desired location in the document.

More complex user interfaces allow more flexibility and support more features, but they can also be confusing and frustrating for users. In particular, users may end up in an undesired state without knowing how to recover. Therefore, it is expected that more advanced electronic document applications will require more training overall. While users may be computer savvy outside of the flight deck environment, their experiences may not necessarily help them in using the EFB electronic documents. Therefore, techniques for manipulating electronic documents should be demonstrated to all users during initial training.

Examples

Electronic documents may have a table of contents similar to the paper document. The user could use the table of contents to see the similarities and differences between the electronic document and paper document.

More sophisticated functionality will require sufficient training to enable users to benefit from the functionality and protect them from getting distracted or frustrated. For example, users will need to understand how to configure documents (e.g., change font sizes, or zoom/pan) if that is allowed. More complex navigation and search techniques will also add to the training curriculum.

Evaluation Questions

- Does the training program provide adequate instruction in how to display and move through an electronic document?
- Does the training program adequately cover how to use the advanced features of the electronic document? Do users know how to avoid using advanced features if they so choose?

3.3 Layout/Appearance

3.3.1 Visual Layout and Structure

Equipment Recommendations

> - Windows and frames, if implemented, should be placed and used consistently.
>
>   NOTE: Frames are selectable regions of the display that may support scrolling, but cannot be repositioned or resized individually like windows.
>
> - Sections of text should be separated using ample white space.
>
> - Where possible, content should be formatted into short segments, each of which communicates one clear point.

Equipment Design Tradeoff(s)

> The visual layout and structure of a document can significantly affect its readability and ease of comprehension. Factors such as font choice, text length, spatial organization, and amount of white space all affect the visual structure of the document. Designers may want to base the visual structure and layout of an electronic manual on the hard copy version of that manual, but the two do not have to be identical. In fact, some aspects of the hard-copy visual layout (e.g., font style and size) may need to be modified for electronic displays.

Problem Statement

Electronic documents need to be designed for ease of reading. Readability is affected not only by how legible the individual characters or words are, but also by how well the visual layout and structure match and reinforces the information structure of the concepts in the material. Use of frames, plenty of white space, and short text segments can all contribute to readability of electronic documents. Note: Visual layout and structure pertain to how text is formatted and arranged on the screen.

Examples

Short text segments are easier to comprehend than long text segments. Long segments of plain text are difficult to read, even if the font is well chosen. This visual structure should match the logical structure of the document in that each short text segment should be focused on communicating one clear point. Ideally, each text segment would be visible in its entirety within the display area of the EFB, so that the user does not have to access off-screen text to comprehend the point.

Portable Document Format (PDF) documents, which capture the visual layout of the printed page, may need to be optimized for electronic presentation rather than just copying the paper version. For example, font choices may need to be modified for legibility on an electronic display.

Use of frames can structure how the user reads the document and improve readability if frames are used consistently. For example, consider the case where the screen is divided into three frames. The upper half of the screen is one frame, and the lower half is divided into two equal sized frames. The top frame could be used to always display the main text of the document. The lower left frame could display a list of related links for navigation, and the lower right frame could always show related figures and/or tables (i.e., detail information).

If the table of contents appears in its own frame at all times, that frame should always be in the same corner of the screen, regardless of the document that is active.

Evaluation Questions

- Does the visual structure of the electronic document match and reinforce the information structure of the document? Is white space used to separate sections of text? Are text segments short?
- Are windows and frames (if implemented) used consistently?

Electronic Documents 79

3.3.2 Minimum Display Area and Resolution

Equipment Recommendations

> - The manufacturer of an electronic document application should identify and specify the minimum display area and resolution necessary to view documents; these two display qualities affect the selection of fonts and the design of the document visual structure.
> - Operators should meet the manufacturer-specified display area and resolution requirements for both training and operational use of the electronic document application.

Problem Statement

In constructing the visual structure of an electronic document, the designer has to start with assumptions about the display area and resolution. Resolution influences the font selection because it affects the visual angle (i.e., "size") of the characters (see 2.4.11). Minimum display area in combination with resolution determines the minimum number of characters per line. Taken together, display area and resolution determine the minimum content that can appear on a screen at one time.

While the visual structure may transfer to a larger content area without difficulty, it is not likely to work properly on a smaller content area that violates design assumptions.

Examples

If an operator intends to use the same electronic document software on EFBs with different content areas and/or display resolutions, they should ensure that all the different EFBs meet the minimum display area and resolution requirements suggested by the manufacturer of the electronic document application. Electronic documents that are visually structured for a smaller display may transfer over to a larger display without difficulty, but the opposite may not be true.

Evaluation Questions

- Does the manufacturer of an electronic document application specify a minimum display area and resolution?
- Does the EFB display intended for this application meet the minimum display requirements?

3.3.3 Off-Screen Text

Equipment Requirement

> - If the document segment is not visible in its entirety in the available display area, such as during zoom and pan operations, the existence of off-screen content should be clearly indicated in a consistent way. (AC 120-76A, 10.b (7))

Equipment Recommendations

> - For some intended functions, it may be unacceptable for certain segments not to be visible. This should be evaluated based upon the application and intended operational function. (AC 120-76A, 10.b (7))
>
> - If part of the document segment is off-screen, the following information should be continuously available to the user:
>
>   (a) how long the document segment is
>
>   (b) how far in the document segment the currently displayed information is

Problem Statement

Documents that will be on an EFB are lengthy and complex. They can be broken down into natural segments, such as different sections and subsections, but it is still possible that a single document segment will be too long to be displayed in its entirety within the available display area. Users then should be made aware of the existence of off-screen content. Plus, the software should be designed to assist users in managing what text is in view and what text is out of view. If they are not aware of off-screen content, users may miss important sections of the document.

With a paper document, the pilot can look ahead to check its length and they can browse or scan the entire document to orient themselves. On an electronic display, it is useful to convey the length of the document, either graphically or numerically. One way of helping the user to orient him/herself is to convey how far in the document the currently displayed information is.

Examples

If the document is implemented in terms of discrete pages, then the current page and the total number of pages can be indicated using a convention such as "1/3," where the first number is the current page, and the last number is the total number of pages. Arrow buttons can also be used to indicate whether there are more pages preceding or following the page currently in view.

If the document is implemented on a scrolling window, a side scroll bar can convey all the required information. A graphical box or bar would represent the location of the currently displayed text in relation to the length of the document.

More sophisticated displays, such as an outline view, can also be very helpful for user orientation, especially for long documents.

Evaluation Questions

- Is the existence of off-screen text indicated clearly? Is the existence of off-screen text indicated in a consistent way?
- Does the software indicate how long the current document segment is and the position of the currently displayed information relative to the entire length of the segment?

3.3.4 Active Regions

Equipment Requirement

> NOTE: Active regions are regions (e.g., selected text, a graphic image, a window, frame, or other document object) to which special user commands will apply, (e.g., hyperlinks or copying)..
>
> - Active regions should be clearly indicated, e.g., through visually highlighting. (AC 120-76A, 10.b (8))

Training/Procedures Requirement

> - Users should know how to activate and deactivate regions.

Training/Procedures Recommendation

> - Users should know the basic special commands that are available for different types of active regions.

Problem Statement

It is often necessary to specify an active region to which special commands will be applied. For example, a text string might be selected for copying into a search query, or a window might be activated in order to bring it to the front of other windows on the screen. Active regions are also useful for selecting between frames on a frame-based visual display. The information in the active frame would respond to update commands entered by the user.

While active regions are not a required feature of electronic documents, if they are supported they should be clearly indicated and users should know how to use them. If the user does not know how to use an active region, he/she will have trouble applying special commands to the intended object. If the user does not know that a particular region is active, he/she may enter inappropriate commands and become frustrated when these commands are not processed as expected.

Examples

Active text could be highlighted using reverse video. Some special commands that might apply to active text include copying and deleting. Note that a sophisticated input mechanism, such as pen input, is required for text selection.

Active windows and frames could be highlighted with special borders. Once highlighted, the window would appear in the foreground (if multiple windows are supported). Only the active window would respond to scrolling, repositioning, or resizing commands. A highlighted frame could respond to scrolling commands, but might not be movable, or resizable.

Evaluation Questions

- Are all types of active regions clearly highlighted?
- Are users trained in how to activate and deactivate regions?

3.3.5 Display of High Priority Information

Equipment Suggestion

> - Some parts of an electronic document may contain high priority information that might be accessed during high workload phases of flight. Because the consequences of user error may be more significant under these conditions, designers should apply more conservative standards for legibility and readability to high priority information, such as a larger typeface and more spacing between lines.

Problem Statement

Although electronic documents will primarily be used during low-workload conditions, it is possible that some parts can be expected to be used during high workload phases of flight, which hold a greater potential for human error. The contents of electronic documents could be reviewed to identify those sections that are more likely to be accessed under conditions that increase the potential for error, the occurrence of which may produce more significant consequences. If more generous standards for legibility and readability are applied to high priority information, the potential for error could be reduced.

Examples

The limitations section, which may be found in the Flight Standards Manual, is one example of a section that might benefit from the application of visual structure that reduces the amount of information displayed at one time. A larger typeface, more spacing between lines, and the use of additional white space between chunks of information could reduce the risk of misreading key information.

Another example is abnormal and emergency procedures described in the Pilot's Operating Handbook or Flight Operations Manual. Appropriate structuring may make the information that the pilot needs urgently, such as a diagnostic procedure, stand out for immediate access and recognition.

There may also be more sophisticated ways of addressing this issue. For example, an electronic document application that is integrated with other flight deck systems could dynamically highlight relevant information on the document.

Evaluation Questions

- Are more generous legibility and readability standards applied to high priority information in the electronic document?

3.3.6 Figures

Equipment Requirements

> - At a minimum, the electronic version of a figure should be able to display all of the content of the paper version.
> - The user should be able to view the overall figure at one time, even if not all the details are readable, in order to get an overview of the figure.
> - The user should be able to read all the details in the figure, although not all of the figure may be visible when the details are readable.

Equipment Recommendations

> - Depending upon the size and complexity of the figure, and the available display area and resolution, the user may have to manipulate the figure to (a) bring areas of the figure that are out of view into view by panning, or (b) make readable details of the figure that are not readable otherwise by zooming. The additional workload of manipulating figures (zooming and panning) is undesirable, so figures should be displayed in their entirety with all details readable whenever possible.
> - Each figure should have descriptive text information associated with it. This text should be available even if the figure is not displayed in full, but marked by a placeholder.
> - Some figures may fit on the display better in either portrait or landscape mode. The user should be able to configure the figure for optimal viewing in either case.
> - If zooming is supported
>
>   (a) discrete zoom levels should be available (e.g. view whole page).
>
>   (b) The current zoom level should be displayed at all times.

Equipment Design Tradeoff

> Figures could be designed to take advantage of the electronic medium in many ways. However, more flexibility in the manipulation of figures increases the complexity of using the software, which in turn impacts user training needs. Also, more flexibility in the manipulation of figures may actually distract users from understanding the actual content in the figure.
>
> Advanced electronic figures may contain complex, variable, dynamic, and/or user-customizable data. These figures need to be evaluated especially carefully to ensure that there are appropriate means of catching any potential errors in the data.

Problem Statement

Figures can be used for various purposes. Graphs are a type of figure. They illustrate relationships between variables. Graphs can also be used as a source for data for calculations. Other figures are representations of relationships between components of systems. Clearly, the usability of figures needs to be ensured in electronic documents.

In paper form, figures are drawn so as to be usable for one or more purposes. The user may be interested in the detail information in the graphic, the overall schematic information in the graphic, or both. The electronic version of the figure has to capture all the content of the paper version, which includes both the details and the overall schematic information. However, there may also be additional workload and training needs that arise from the need to configure electronic figures for maximum usability. Designers should consider these tradeoffs when implementing more complex features for electronic figures.

Examples

In order to load data more quickly, the figure may not be displayed until specifically called up. The associated text information should describe the contents of the figure so that the user knows whether they want to call up the actual figure.

Some ways in which figures could take advantage of the electronic medium include interactivity and customization. For example, figures could be interactive in that the user could specify which "layers" of the drawing they wanted to be visible or not, in order to understand complex relationships. A graph could also be interactive in that the user could select parameters for producing the plot.

An electronic document application that was integrated with other flight deck systems could also produce custom graphs or figures relevant to the current situation. For example, if the weight of the aircraft is known, graphs and figures could show data that apply for that aircraft configuration.

The descriptive text associated with a figure need not be part of the document itself. The text could be provided by naming the electronic figure, where the name is accessible by hovering over the figure, or by selecting the figure.

Evaluation Questions
- Can the user view the entire figure at one time?
- Can the user easily configure the display such that details of the figure are readable?

3.3.7 Tables

Equipment Design Tradeoff(s)

> Because of screen resolution limitations, translating tables from paper to electronic format may require modifications to their design to ensure that users can quickly locate target information contained in the tables.

Problem Statement

Tables typically provide a lot of data in a relatively compressed space. Borders and lines are often used to help the reader visually organize the information correctly. The comparatively low resolution of EFB screens, as compared with paper, may require that the design of existing tables be reconsidered to ensure that comparable readability is achieved.

Table size may also need to be reconsidered. Paper tables may utilize the full height and width of a standard page size. EFB screen size may be significantly smaller. A larger type size may also be required, which may require rethinking of how much data can be supported by an individual table.

Examples

Additional white space may be required to clearly separate individual table elements from each other and also from neighboring borders or lines. Placing column and row names in bold can help the user to interpret table information more efficiently.

If the electronic document application is aware of the aircraft systems' status, it may be possible to customize and reduce the information in a table such that only the relevant data are displayed. For example, if the weight of the aircraft is known, or the ambient temperature is known, data that apply only to those and similar conditions could be displayed.

Evaluation Questions

- Are the tables included in the electronic documents as readable and as usable as their paper counterparts?

3.4 Navigation and Searching

3.4.1 Moving to Specific Locations

Equipment Recommendations

- If there is a cursor, it should be visible on the screen at all times while in use. (AC 120-76A, 10.b (7))
- If the electronic document application supports links, entries in the document table of contents and indices should be linked to the corresponding locations in the text. Cross-references should also be linked to each other within a document.
- The user should be able to cancel a movement by returning to previous location in one step.

Equipment Suggestions

- The electronic document application should track the most recently visited sections in the document and allow the user to select from this list to return quickly to a recent location. The resulting action for relative movements (e.g., "back" or "forward") should be predictable.
- Cross-references should be linked across documents if possible.
- If the electronic document application supports end-user customization, users should be able to configure and manage their own bookmarks to selected locations in the text.

Problem Statement

The manuals that pilots use are typically lengthy documents. It is important that users be able to navigate quickly to important locations within these lengthy manuals. Key locations in the document include the beginnings of each new section, which are listed in the table of contents, cross-references, and index entries. Each of these types of information can and should be linked to the appropriate location in the text. Other key locations in the manuals include recently visited locations and entries in the indices.

Because users often find themselves revisiting certain parts of an electronic document, it is also desirable to implement customizable electronic "bookmarks." These bookmarks would be set and managed (e.g., renamed or deleted) by the user, similar to managing favorite links on a web browser.

Sometimes users will make a mistake and unintentionally move to a location. For these cases, the user should be allowed to cancel the movement and return to their previous location in one step.

Examples

Hyperlinks to different locations in the text are one way to move about a lengthy document quickly. Another way of moving about the document could be by selecting a location to move to from a list. For example, the last five visited locations could all be listed in one place (a "catalog" page), and the user could select which of these locations to move to by selecting one from the list. The selection need not be done by a pointing device. Selection could be accomplished through soft keys, which change the selected location.

Evaluation Questions

- Is the cursor easily visible when in use?
- Are the table of contents and indices linked to the corresponding locations in the text?
- Can the users quickly return to recently visited locations in the text?
- Are recently visited locations tracked so that the user can return to them quickly?

Electronic Documents 87

3.4.2 Managing Multiple Open Documents

Equipment Requirements

> NOTE: The active document is the one that is currently displayed and responds to user actions.
>
> - If the electronic document application supports multiple open documents, the system should indicate which document is active, and display that indication (e.g., its title) continuously. (AC 120-76A, 10.b (9))
> - Under non-emergency, normal operations, the user should be able to choose which of the open documents is currently active. (AC 120-76A, 10.b (9))

Equipment Recommendation

> - If the electronic document application supports multiple open documents, a master list of all open documents should be available.

Equipment Suggestions

> - Access to a document can be provided through the master list of open documents.
> - If the display area is large enough, it is useful to be able to arrange multiple open documents such that text from more than one document is visible at the same time.

Problem Statement

If the electronic document application can support multiple open documents, the user will need assistance in managing these documents, or else they may become confused about what document they are using because the visual structures of the documents may be very similar. In particular, the user has to keep track of which documents are open, and which of these is the currently active document. Document titles help the user manage multiple open documents. The user should also be able to move between the open documents quickly. That is, the user should be able to activate an open document from a list of open documents, rather than from a list of all available documents.

Examples

If the electronic documents are running on an industry standard windows-based GUI, document titles should appear at the top of the window frame by convention. Different documents could be selected by activating the particular window that contains the document of interest.

The user may want to open both the POH and the MEL in order to see some cross-referenced items. He/she may want to arrange these two documents such that the top half of the screen shows one document, and the bottom half shows text from the other.

Evaluation Questions

- If multiple open documents are supported, is the title of the active document shown continuously? Can the user easily choose which open document is active?
- Is a master list of open documents available?

3.4.3 Searching

Equipment Recommendations

- The electronic document application should support multiple search techniques. Some options include searching by:

    (a) user-entered key words and their combinations

    (b) links to text (e.g., via cross-references or a table of contents)

    (c) graphical links (e.g., look up the function of a switch based on its location in the flight deck)

    (d) header/footer information (e.g., chapter and/or section titles written at the top of each page)

- Users should be able to select which document(s) the search should cover.

Equipment Design Tradeoff(s)

If key word search is implemented, the EFB may need to support entry of free text. Also, designers will have to consider the design of a query language, e.g., support of terms such as AND/OR. Complex query languages may have training implications. Also, designers will have to consider how the user will move between multiple hits from a key word search.

Problem Statement

One of the advantages that electronic documents can have over paper documents is the ability to search a lengthy document quickly. With paper documents, search is conducted through section headings, or indices. There are other options that can be implemented electronically, such as search based on user-entered key words, and text and graphical links. These electronic search techniques can be quite effective when the user is conducting a well specified search. Browsing header/footer information can be effective when the object of the user's search is not specified well enough for other types of searches.

Examples

To find definitions for various types of runway surface conditions, the user could do a search on the key words "runway," "surface," and "definition," or he/she could search via links from the table of contents of the flight operations manual.

The pilot may need to access the MEL just before departure. He/she could initiate a search based on the name of the faulty equipment.

Evaluation Questions

- Are multiple search techniques supported in the electronic document application?

3.5 Options

3.5.1 Printing

Equipment Requirements

> - The electronic document application should clearly indicate which pages or document sections have been selected for printing.
> - The user should be able to immediately terminate a printing session for any reason.

Equipment Recommendations

> - Users should be able to select document subsets for printing, including individual sections and individual pages.
> - The visual structure of the printed document should match the visual structure of the electronic document displayed on the EFB. (The printed document does not need to be identical to the original paper document.)

Training/Procedures Suggestion

> - If flight deck printing technology is limited in capability and cannot meet the recommendations above, users should be aware of these limitations.

Problem Statement

Users may wish to have a hard copy of some portion of an electronic document in the flight deck. To date, printing technology for the flight deck is expensive and it is often much less capable than typical office printing technology. However, if it is available, the flight deck printing option should enable users to select, as precisely as possible, the subset of information of interest so that the user is not inundated with irrelevant material. In addition, users should be able to terminate a long print job in the event that an error has been made, so that the user can quickly return to the application to print the correct section. The visual structure of the output should also match what is displayed on the EFB. If it is not possible to match the output with the EFB display, users should be aware of the printing limitations.

Examples

The printing option should allow users to choose between printing either a page range (including a single page) or one or more sections. The print window should clearly indicate the print range that has been selected. To help ensure that the correct pages have been selected for printing, the printed version should closely correspond to the electronic version in that the subset of information that is printed is identical to the information displayed electronically.

In the event that an incorrect page range is selected, the user should be able to immediately terminate the printing session if more than a few pages have been selected for printing. Otherwise, the printer will be temporarily unavailable for further use and paper will be wasted. There are many other reasons why a print job may need to be terminated. Even a printout of a single page may need to be cleared or terminated.

Evaluation Questions

- Does the EFB allow users to choose the subset of information to be printed?
- Does the printed version correspond to the electronic version displayed on the EFB?
- Does the EFB clearly specify the subset of information that has been selected for printing?
- Can the user immediately terminate a print job?

3.5.2 Animation

Equipment Requirements

> - If animation is supported, the user should be able to start and stop the segment. The user should be able to stop the animation at any time, even if the segment has not ended. (See also 2.4.14 if the animation has associated audio.)
> - There should be supporting text to describe and support the animation. This text should be available even if the animation is not currently running.

Equipment Recommendations

> - Animation should only be used to highlight and explain important relationships. It should not be overused.
> - If the animation has associated supplemental audio, control of both the audio and video should be integrated.

Problem Statement

Animation can be a powerful aid to visualization of complex relationships. It is especially useful for training or study of new or detailed material. In these situations, the user should focus attention on the animation for it to be of value.

Because of the need to focus attention on complex animations, its use in flight may need to be limited. At the very least, the user should be able to interrupt any animation in progress, in order to quickly switch to higher priority tasks. Also, supporting text should be available to preview the content of the animation. In other words, the user should not be required to start the animation in order to determine its contents.

While animation can be beneficial for specific purposes, overuse of animation should be discouraged. Too much visual movement on the screen can distract the crew from higher priority tasks.

Examples

Animation could be implemented to show how parts of a complex mechanical system fit together, or as part of a training tutorial on use of the software.

Animation should not be used to highlight company logos.

Evaluation Questions

- Can the user control when the animation begins? Is the animation interruptible?
- Is there supporting text for the animation that identifies its contents without running the segment?

Electronic Documents   91

3.5.3 Making Notes

Equipment Design Tradeoff(s)

> Paper documents are customizable in that users can make notes and highlight information of interest to them. Adding these customization features to the electronic document functionality could also be useful (although it is unlikely to replace pen and paper entirely). The utility of a notes feature may be constrained if users do not have access to their own notes while using an EFB. To ensure that the users can view their own notes, they may have to enter the notes on a personal EFB, or their notes may have to be accessible through a server that communicates with all EFBs.

Problem Statement

One of the advantages of paper documents is the ability for the user to underline, scribble notes in the margins, and otherwise utilize the document to record their own information. Users would benefit from having a similar capability with the electronic document function. However, the EFB may not be a personal unit, in which case, the notes may not be very useful to the other users.

The extent to which a notes feature would be useful will be determined in part by the company policy on EFB use. EFBs that are used only on the flight deck may benefit less from this feature in that users would have access to these notes only while on that specific aircraft.

Examples

The electronic document function could provide the capability to select chunks of text that could be highlighted in a fashion similar to underlining in a paper document. Similarly, a location in the text could be selected and a notes feature utilized to allow the user to record their own notes. Retrieval of notes could take place in two ways. First, access to the set of all notes that have been recorded could be supported. Second, a visual indicator could be placed in the location from which the note was created which would then be used to access the individual note that was created for that location. It is possible that these notes could be stored in a central ground-based server that would allow upload to the specific EFB that will be used by the note creator for each flight. If the server is used to store other information "owned" by that user, this might be useful to do.

Evaluation Questions

- If the electronic documentation supports note taking, can users always access their personal notes?

3.5.4 Decision Aid/Automatic Call-up of Data

Equipment Design Tradeoff(s)

> The integration of an EFB with other flight deck systems may enable electronic documents to serve as a decision aid. Consideration should be given to integrating electronic documents into systems that sense aircraft status to provide users with immediate access to information that can support more effective flight management.
>
> Although decision aids can have value in supporting more effective decision-making, they can also have the unintended consequence of excessive reliance, where the crew may become complacent about reviewing the suggestions made by the system.
>
> Note that, if the EFB is used as a decision aid, its software and hardware may need to meet more stringent reliability and availability criteria than otherwise. A formal safety analysis may need to be conducted.

Problem Statement

Integrating the EFB with other flight deck systems could allow the electronic documentation application to customize its information based upon current flight conditions. Doing so can help to reduce crew workload by reducing the amount of information the crew has to consider. An unintended consequence can be complacency, in which the crew relies on the decision aid to select information for review without sufficient crew involvement.

Examples

If the electronic document application is aware of aircraft system status, its tables and other content could be customized for the current situation automatically. For example, knowing the fuel burn rate would enable the decision aid to list only the candidate airports within range if the crew needed to divert to a different airport.

Several approaches can be taken to mitigate the problem of complacency. First, the decision aid could offer several options rather than a single answer; doing so would encourage the crew to review the information needed to select an option. A second approach is to require the user to make the decision first, then have the decision aid review the soundness of that decision.

Evaluation Questions

- Does the decision-aiding function mitigate the risks of crew complacency?

4 Electronic Checklist Systems

Checklists are used by pilots to ensure that critical actions are completed at appropriate times. Paper checklists are in common use across all operations, but electronic checklist (ECL) systems are increasingly common, particularly on newer business jets and air transport aircraft (see Appendix A).

ECL systems are a logical fit for EFBs; normal, abnormal, and emergency checklists are specifically called out in AC 120-76A as a Type B application. The guidance for ECL systems in the EFB AC references FAA AC 120-64 (Operational Use and Modification of Electronic Checklists). However, the focus of AC 120-64 is on high-end ECL systems, such as the system on the Boeing 777, and its guidance applies only to Part 121 and Part 135 operations. For this report, we assume that paper backups of the electronic checklists are available, as specified by AC 120-64, and as recommended elsewhere (CAP 708).

Background information on checklists is provided below, in Section 4.1. General guidance on the human factors considerations associated with EFB ECL systems is provided in Section 4.2. Guidance on interacting with checklists is provided in Section 4.3, and guidance on interacting with items within a checklist is provided in Section 4.4. Finally, in Section 4.5, guidance on some more advanced and unusual ECL system features is provided.

4.1 Background

4.1.1 Research on Paper and Electronic Checklists

There is a significant body of well-documented research that has been performed on paper and electronic checklists (see References). Much of the research pertains to the design and use of paper checklists (see FAA, 1996 for a review of that literature). Design issues such as language, content, and formatting all affect overall ease and accuracy of using checklists. These issues transfer over by and large to electronic checklists, and are not discussed in this report. Much of the published research on ECL systems was performed by the Boeing Company during its development of the 777 ECL system (see Boorman 2000 and 2001). The British Civil Aviation Authority also published a thorough report (CAP 708) that addresses ECL issues for air transport operations. CAP 708 also provides a good overview of checklist design issues.

Use of paper checklists is prone to various types of errors, which have been associated with numerous accidents (CAP 708, Boorman 2000, 2001). ECL systems could, in theory, mitigate errors made with paper charts. However, Boorman notes that ECL systems could also introduce different types of errors, and so must be designed with care. One potential hazard of ECL systems is the compelling nature of the display. Crews may believe that the ECL has more intelligence than it really does, leading them to treat the information less critically than the paper checklist. For example, Boorman (2001) states that specific coaching is sometimes required to remind the captain that he/she is the final judge as to which procedures to carry out and when.

The guidance on ECL systems provided herein matches well with other literature on ECL systems. The main difference is that this chapter also provides guidance on human factors considerations for ECL system that may be much simpler and less sophisticated than the highly automated systems discussed in AC 120-64, CAP 708 and Boorman (2000, 2001).

4.1.2 Features of ECL Systems

Paper and ECL systems are compared and contrasted along several dimensions in Table 4-1 and Table 4-2. For some dimensions, paper is compared against generic ECL systems, but for the other dimensions, we distinguish between *simple* ECL systems, *advanced* ECL systems, and ECL systems that can *sense aircraft state* and/or has connectivity with aircraft systems.

A simple ECL system has an underlying database of items and checklists, and features that make the item status distinguishable (i.e., memory aids), but the application does not try to trap pilot errors. A simple spreadsheet application (such as a list-generating program) could suffice for this type of ECL. While simple ECL systems are not necessarily an improvement over paper checklists, and in fact could be more difficult to use than paper checklists, we do consider them here because they are available in the EFB market. The simplest possible ECL would be a non-interactive electronic display such as a PDF version of a paper checklist. While this meets the basic definition, and possibly even meets the functionality of paper checklists, it is not likely to be a popular option because of its limited functionality and benefits. Instead, it is expected that most ECL systems will at least be based on a database of checklists and checklist items.

An advanced ECL system implements features that help to trap pilot errors. There are a variety of checks that could be implemented to safeguard against common errors made when using paper checklists (e.g., the crew fails to initiate or complete the checklist, or the crew overlooks items on the checklists).

ECL systems that sense aircraft state are even more sophisticated in that some of the items may marked as completed through sensing rather than pilot input. For example, the ECL system could sense the position of a switch, and automatically mark the item complete if the switch is in the correct position. Note that checklists that are "interactive with other aircraft systems" are specifically excluded from a list of Type B applications in AC 120-76A. However, we include them here because they may be part of fully certified EFB systems.

Physical Display	Paper Checklists	
	Paper checklists can be manipulated easily and (when laminated) are robust enough for daily and emergency use.	
	ECL System	
	If the EFB is installed, the workload of locating and positioning the paper checklist is eliminated.	
	If the EFB is not mounted or installed, it may need to be stowed for high workload phases of flight, making even normal checklists unavailable (e.g., for takeoff and landing).	
	If high-quality electronic display, the checklist may be more readable than paper under a range of lighting conditions, but low-quality electronic display could be less readable than paper.	
	If the display is small or has low resolution, may not be able to show the whole checklist at one time.	
Legibility and Understandability	Paper Checklists	
	Information structure and visual layout is highly familiar and has been designed for ease of use.	
	ECL System	
	Legibility is affected by display quality, size and resolution. Understandability is affected by familiarity and complexity of formatting.	
Accessibility of Individual Checklists	Paper Checklists	
	Pilot has to locate the paper checklists and manually search within a set of checklists for the particular checklist of interest. This task may be easier or harder depending upon the physical form of the paper checklists (e.g., bound versus separate sheets).	
	Simple ECL System	
	Could support multiple methods of searching for a checklist. For example, user could select from a master list (menu) of checklists, possibly even a hierarchical menu, or user could search for a checklist using key word search. However, poorly designed methods of access may require several steps, increasing workload.	
	Advanced ECL System	
	System could suggest checklists in normal sequence of use to lead crew through all the checklists they need to complete, helping to ensure that all necessary checklists are initiated. Flexibility should be maintained though; the crew should still be able to view checklists in a different order.	
	ECL System with Sensing	
	Checklist could be suggested or brought up automatically upon detection of a condition (e.g., a fault condition, or flight stage, such as passing 18,000 ft during climb out) to reduce workload of selecting and accessing appropriate checklists, especially for non-normal conditions. Crew may need to be aware of alternatives to avoid over-reliance on automation.	
Links	Paper Checklists	
	All links are tracked mentally by the pilot.	
	ECL System	
	Links to supplemental material or calculation worksheets could provide more direct access to related information. But, for simple checklists, this may be an unnecessary feature that increases complexity of the user interface.	
	Could also provide more direct access to related *checklists*, which is especially useful for branched or linked checklists (e.g., embedded checklists for non-normal procedures).	

Table 4-1. Comparison of paper and electronic checklist features (continued on next page).

Within-Checklist Tracking	Paper Checklists	The pilot tracks item and checklist status manually and/or mentally. Error modes include skipping items inadvertently and forgetting to complete checklists.
	Simple ECL System	System tracks item completion and gives crew a positive indication of which checklist items have been completed. Crew entry is required, potentially increasing workload and time to complete the item.
		Ability to indicate the status of individual checklist items. Provide item coding as a memory aid, e.g., completed items formatted differently, or have a symbol next to them, such as a checkmark. Also, item grouping could be varied manually (e.g., show all items not yet completed). However, complex item codes (e.g., with many colors and symbols) could be confusing and make the display less usable.
	Advanced ECL System	System tracks and indicates active item to serve as a placeholder and memory aid.
		System tracks skipped, deferred, or overridden items to give crew a positive indication of checklist items that have been reviewed and are either still pending, or have been overridden by the crew.
		System tracks whether all items on the list have been completed; generates "Checklist Complete" message when appropriate to prevent the crew from closing an uncompleted checklist inadvertently. Should allow crew to exit checklist with confirmation of pending items.
	ECL System with Sensing	Item status could be verified or completed automatically by sensing aircraft state or switch position. Manual completion of sensed items would not be necessary, reducing workload. However, there is a potential for crew to be unaware of the status of items that are completed through sensing aircraft state or switch position.
Managing Multiple Checklists	Paper Checklists	Pilot is responsible for managing checklists manually.
	Simple ECL System	System may allow toggling between open checklists through a master list.
	Advanced ECL System	System tracks and indicates active item to serve as a placeholder and memory aid. Especially useful if multiple checklists are open.
		System tracks which checklists have been completed to help prevent skipping a checklist inadvertently.
		System tracks and indicates status of individual checklists if multiple checklists are open (completed vs. pending) to help remind the crew of the next checklist(s) they need to accomplish when many checklists are open.
	ECL System with Sensing	Checklist status could be verified or marked as completed by sensing aircraft state or switch position. Manual completion of sensed checklists may not be necessary, reducing workload. However, there is a potential for crew to be unaware of the status of checklists that are completed through sensing aircraft state or switch position.
		When checklists are called up as a result of a sensed condition, they can be prioritized. For example, a non-normal checklist would supersede a normal checklist if both are called up as a result of a sensed state.
Customization	Paper Checklists	Pilot has to track custom elements (e.g., as checklists branches) mentally.
	Simple or Advanced ECL System	Manual input can help to customize conditional checklists. For example, the checklist could be customized for a particular system (e.g., left or right engine) eliminating the need for the crew to recall which system is affected.
	ECL System with Sensing	Sensed information can help to customize conditional checklists.
Reminders	Paper Checklists	None.
	Advanced or Sensed ECL System	System could generate reminders to complete skipped, delayed, or deferred items at appropriate times to help ensure that all items are reviewed.

Table 4-2. Comparison of paper and electronic checklist features (continued from previous page).

4.2 General

4.2.1   Checklists Supported by the ECL System

Equipment Requirements

> - If the ECL system supports normal checklists, then *all* normal checklists should be supported.
> - Similarly, if the ECL system supports non-normal/emergency checklists, then *all* non-normal/emergency checklists should be supported.
> - Similar requirements apply for other categories of checklists if applicable (e.g., supplementals.)

Equipment Recommendations

> - If an ECL requires access to a checklist that is not supported by the system, the ECL system should indicate the location of the unsupported (i.e., paper) checklists.
> - Non-normal checklists supported by the system should retain as much commonality with normal checklists as possible so that pilots are familiar with the user interface even if they encounter an unfamiliar checklist.

Equipment and Training/Procedures Recommendation

> - Checklists that may need to be used in high workload conditions (e.g., non-normal procedures) should only be used on installed or mounted EFB hardware (instead of unsecured hardware) to reduce crew workload.

Training/Procedure Requirement

> - Flight crews should know which checklists the ECL supports.

Problem Statement

If the ECL system does not support all the checklists available to the flight crew, the crew has to keep track of where to find a particular checklist. In order to make this task easier for the crew, the ECL should at least support all checklists within categories that are meaningful to the flight crew. In this way, the crew need only recall that a certain category of checklists is supported either in the ECL system, or on paper, helping them to locate checklists more quickly and accurately. An ECL system that does not include all of the checklists that belong to a supported category can create confusion and lack of trust in the system when crew members attempt to find a checklist that is not in the application.

Examples

A simple ECL system is likely to contain normal checklists, but may not contain non-normal checklists. If a normal pre-takeoff checklist is in the ECL, then the crew would reasonably expect to see other normal checklists in the system, such as those for takeoff, climb, descent, and approach/landing.

Evaluation Questions

- What types of checklists does the ECL system support? Are all checklists within those categories available on the ECL?
- Is the location of the paper checklist provided in all electronic checklists that require subsequent access to an unsupported checklist?
- Do pilots know which checklists are supported electronically and which are not?

Electronic Checklist Systems 97

4.2.2 Information and Visual Layout/Structure of Electronic Checklists

Equipment Requirement

> - The paper and electronic versions of the checklist may not be identical; however, the resultant crew actions called for in the checklist should be the same regardless of the version in use. (AC 120-64)

Equipment Recommendations

> - Layout of items on an ECL system should be based on the same layout as the equivalent paper version. Line length and page size may differ (e.g., because of the electronic screen may be smaller than the paper), but the use of headings, sub-headings and titles should be consistent. (CAP 708)
> - Similar to paper checklists, the format of the electronic checklist should make it clear which challenge is associated with which response (e.g., by connecting the two with a dotted line). (CAP 708)
>
> NOTE: The challenge is normally the name of the system or control or parameter involved in the action, and the response is normally the required status of the system or control or parameter (e.g., challenge is "approach speed" and response is "Vref + 10").

Problem Statement

Paper and electronic checklists must both be completed accurately to ensure safe operation of the aircraft; therefore any differences in formatting/appearance should not translate into differences in crew actions.

The formatting of paper checklists is designed to aid the pilot in checking and completing items accurately and quickly. Pilots are also highly familiar with the formatting used in paper checklists. For both these reasons, it is recommended that the visual appearance of ECL be consistent with the appearance of paper checklists. If the format of the ECL is radically different and untested, there is the possibility that crew may have difficulty in using it quickly and accurately. Also, if the ECL system does not replace all paper checklists, and the crew may be expected to use a paper checklist now and then, it is especially important that the electronic and paper formats be consistent, so that crews do not make errors in using one or the other.

Examples

CAP 708 contains several recommendations for formatting and layout of ECLs. For example, each item on the checklist should be contained on one line of the screen. If the item is longer than one line of text, the spacing between that item and the next should be greater than the spacing between lines within the item.

Evaluation Questions

- Are the resultant crew actions the same regardless of the format (paper or electronic)?
- Is the layout and formatting of the ECL clear?
- Is the layout and formatting consistent with the paper checklist equivalent?

4.2.3 Checklist Data and its Modification

Equipment and Installation Requirement

> - Checklist data (e.g., titles, item text, sequence of items, etc.) should be current for the aircraft (see AC 25.11 for more specific guidance).
>
> NOTE: See AC 120-64 for further definition of checklist data and guidance on Part 121/135 procedures for operator modification of ECL data. Part 91 operators should also review this guidance to understand the issues involved.

Equipment and Training/Procedures Recommendations

> - Paper deviations and supplements should not be required to keep ECL data current.
> - ECL data should be easily modifiable by the operator or by an operator-selected systems integrator, rather than requiring modification by the manufacturer.
> - Procedures and tools developed for modifying checklists should minimize the opportunity for introducing errors.
> - The modification process for ECL data should also ensure that conflicting versions of the ECL are not in use.
>
> NOTE: Procedures and tools for modification of ECL data may be subject to regulatory review and approval in order to ensure that modifications are made accurately, particularly if the ECL supports checklists items that are sensed. See AC 120-64 for guidance on modification of ECL systems for Part 121 and Part 135 operations.

Equipment and Installation Recommendation

> - If aircraft signals are used by the ECL system, the configuration of the ECL database must be carefully controlled to ensure that the signals required for ECL operation are always available from the aircraft and have the correct logic to match the ECL function they support.

Problem Statement

Any ECLs in use (normal or non-normal) need to be accurate and up to date with regard to the aircraft in which they are used. Because checklists (both paper and electronic) may need to be updated when there is a change to the aircraft, and at other times as well, most operators will prefer to manage the content in-house, or through a contracted systems integrator, rather than through the aircraft manufacturer who supplied the original EFB. In particular, it is important that urgent or significant changes to ECLs are made efficiently, in order to avoid use of paper supplements. Paper supplements could be forgotten, and cause confusion and errors, particularly during high workload non-normal conditions.

Examples

Ground-based tools used by the operator could support automatic documentation of changes to ECL data, to construct an audit trail. The audit trail could be used by company and FAA personnel to verify and review changes. The audit trail could record which ECLs were changed (e.g., which fleet was affected) and why. It could also verify that all necessary related changes were made.

If changes to the ECL cannot be made in a timely fashion and a paper supplement is required, the air carrier should consider whether to issue a paper supplement with appropriate notification to flight crews or whether to temporarily return to use of paper checklists instead of the ECL system

Evaluation Questions
- How are modifications to ECL data made? Are ECL data kept current for the aircraft?
- What precautions are in place to avoid errors during changes to the ECL system?
- How are changes to ECL data documented?

4.3 Interacting with Checklists

4.3.1 Accessing Checklists

Equipment Requirement

> - All checklists supported by the ECL system (e.g., normal and non-normal) should be individually accessible for reference or review at any time while the system is active. (CAP 708)

Equipment Recommendations

> - If the ECL application supports normal checklists, they should be accessible in accordance with the normal sequence of use.
> - Accessing electronic checklists should be at least as quick and accurate as accessing paper checklists. (CAP 708)
> - Although the ECL system may keep track of the next checklist to be used, it should only open that checklist upon crew input, not upon completion of the previous checklist.

Equipment and Installation Recommendations

> - When a non-normal condition is detected by a sensing ECL system, the system should alert the crew that a checklist (or more than one checklist) applies to this condition, but it should only call up the appropriate checklist when commanded by the crew to ensure that the crew has fully reviewed the situation and concurs with the applicability of the checklist proposed by the ECL system.
> - If more than one checklist could apply in the above situation, the crew should be given a list of suggestions.

Equipment Suggestion

> - Access to open checklists could be provided through the master list of all open checklists. (See 4.3.2.)

Problem Statement

Normal checklists are used to ensure proper aircraft configuration at key points in each phase of flight, but the crew may also need to review them at any time. Similarly, non-normal checklists may be called up during an actual non-normal situation, or for review.

In some checklist-related accidents/incidents, crews have forgotten to review a normal checklist at the appropriate time. If the ECL system suggests normal checklists in the usual order, this type of error could be trapped.

Accessing checklists needs to be at least as quick as paper because the checklists (especially non-normal checklists) may need to be brought up during high workload situations.

Examples

Normal checklists are typically performed in a fixed order, so the ECL system may be designed to cue the next checklist in the sequence after the completion of its predecessor. With this feature, the crew would not need to locate and access the next checklist under normal conditions.

Access to all individual checklists could be supported from a "table of contents," which could be a menu or list of checklist titles that are hyperlinked to the actual checklists. The organization of the contents list should be consistent with that used in the FAA-approved flight manuals. Non-normal checklists could also be accessed manually through the table of contents. Organizing checklists by the type of non-normal

Electronic Checklist Systems 101

condition, as found in Quick Reference Handbooks, may be appropriate. Formatting the contents list with clear checklist titles, together with a list of the indicators for the corresponding non-normal condition, can help crews to manually select checklists accurately.

Additional schemes for organizing checklists include ordering alphabetically (by title) or by subject.

An ECL could provide a control or menu for moving between open (active) checklists only. Checklists that are closed (not in use) would not be accessible through this mechanism, simplifying access to the most pressing checklists.

Evaluation Questions

- Are normal checklists accessible in the usual order of use?
- Can all checklists be accessed individually for review or reference?
- If the ECL system proposes a checklist to the crew for management of a non-normal condition, is crew input necessary to open up that checklist?

4.3.2 Managing Checklists

Equipment Requirements

- The title of each open checklist should be continuously visible, even if only one checklist can be open at a time. (CAP 708)
- The checklist title should be displayed above the items and be distinguished throughout the checklist (i.e., even when the first item is no longer visible). (CAP 708)
- If the ECL system supports more than one open checklist, the user should be able to access other checklists without first having to close the currently displayed checklist. (See also 2.4.5 on Multi-Tasking, and 3.4.2 on Managing Multiple Open Documents.)
- If more than one checklist can be open, the user should be able to choose which checklist is currently active.

NOTE: The active checklist is one that is currently displayed and, if the ECL system tracks item status, the items on the active checklist change in response to user actions.

- If one checklist is a "child" of another checklist in which it is embedded, the user should be able to choose whether the parent or the child checklist is active.

NOTE: If one checklist is embedded in another, the higher-level checklist is called the "parent," and the embedded (lower-level) checklist is called the "child."

Equipment Recommendations

- If item status is tracked *and* more than one unrelated checklist can be open, or if the EFB supports multiple functions that can interrupt checklist completion, a placeholder capability should be available to remind the user which item was active prior to leaving the checklist.
- There should be a simple and distinct crew input that resets a checklist to the state where all items are marked as incomplete, in case the crew needs to start with a fresh checklist at any time.
- In place of parent-child checklists, the system should use a single checklist that incorporates both.
- If multiple checklists can be open, a master list of all open checklists should be available.

Problem Statement

Working with electronic checklists can be a complex task. At a minimum, the user has to manage which checklist is active and keep track of item completion. If multiple checklists are open, the task complexity increases significantly. To minimize the workload of managing checklists, the ECL system should provide the user with information that can help him/her to manage this task accurately and efficiently.

Examples

The crew is performing the 10,000 foot/climbing checklist. A left cowl anti-ice message appears on the EICAS. Because the cowl anti-ice is on, a low pressure condition is indicated. The climb checklist is not complete but they need to open the cowl anti-ice checklist because icing conditions take priority. They perform the first part of the checklist, which ends with the command to maneuver out of icing conditions. After reconfiguring the aircraft to depart the icing conditions, they return to the climb checklist. Then, they return to the cowl anti-ice checklist which, on the EFB, includes a natural transition to the ice dispersal procedure. Both of the open checklists are titled so that the crew knows which one is in view at any time.

The user should be able to move between both related and unrelated checklists. For example, a checklist with a related (embedded) child checklist should allow the user to access the child checklist. Once the child checklist is active, there should be a way for the user to return to the parent checklist.

Pilots may need to reset all items on certain checklists to incomplete status in the event of a go-around, or touch and go operation.

Evaluation Questions
- Does each checklist have a constantly visible title?
- Can the user move from between open checklists easily?

4.3.3 Managing Non-Normal Checklists

Equipment Recommendations

- When managing multiple non-normal conditions with an ECL system that is aware of aircraft state, all checklists associated with the on-going non-normal conditions should be listed together in one master list, separate from other checklists.
- The master list of checklists to be completed for managing non-normal conditions should indicate the status of each checklist (e.g., pending, accessed/partially complete, complete, or overridden).

Equipment Suggestions

- Access to a specific checklist during a non-normal condition can be supported from the master list of required checklists for that condition.
- Users should easily be able to access procedural, system, and operational notes and other information pertaining to any ongoing non-normal condition.

Equipment Design Tradeoff(s)

Non-normal checklist items that are to be performed at a later time could be automatically integrated into subsequent normal checklists. Integrating these items into later checklists can ensure that pilots do not forget to complete the tasks, even while busy. However, it also adds to the complexity of the ECL system, and more significantly, adds complexity to the operator's task of correctly modifying the checklist database and validating the correct operation of these items. A potentially simpler option would be to prompt the pilot back to the non-normal checklist at the appropriate time to finish the delayed items.

Problem Statement

Managing non-normal conditions can be a high workload task that requires use of more than one checklist. While managing the non-normal conditions, the crew must also keep track of which checklists need to be performed and the status of each checklist, a task that places additional demands on a crew that may already be overloaded. An ECL system that senses aircraft state or switch positions could help the crew manage the performance of all the required checklists while providing crews the flexibility to schedule checklist performance in accordance with other high-priority tasks.

Examples

Providing direct access to checklists through a master list of checklists can reduce the potential for selection of the wrong checklist. This list could be used to cycle through all checklists that have to be completed, including any that were open when the non-normal conditions occurred.

Ongoing non-normal conditions, inoperative systems, procedural changes, operational limitations, and other information could be listed in one area. Advanced versions of this function could allow the user to select which checklists to pull this information from or to choose specific paragraphs from operating manuals and other documents for inclusion.

Evaluation Questions

- Does the ECL indicate the checklists that should be performed when multiple malfunctions have occurred?
- Can the user easily access these checklists?

Electronic Checklist Systems 105

4.3.4 Lengthy Checklists

Equipment Requirement

> NOTE: A multi-screen checklist is one that has more items than can be displayed at one time on the display (i.e., the checklist can not be displayed in its entirety on one page/screen).
>
> - The ECL should allow the user to look ahead in a multi-screen checklist (e.g., page down) without changing the active item.

Equipment Recommendations

> - While a multi-screen checklist is in use, the following information should continuously be available
>
>   (a) How long the whole checklist is
>
>   (b) How far down the checklist the currently displayed information is
>
>   (c) How much of the checklist has been completed
>
> - It should not be possible to change the status of an item if that item is off screen.
> - If the active item is off-screen and the user makes an "item completed" entry, either an error message (e.g., "Active item is off-screen") should appear, or the active item should be called into view.

Design Tradeoff(s)

> Paging and scrolling both give the user access to all items in a lengthy checklist. Paging allows the flight crew to see where they are in the current set of items clearly. There is no unexpected movement of the screen, and an item always appears on the same place on the screen. Scrolling allows the user to view previous items on the checklist and forthcoming items simultaneously. However, scrolling needs more careful management of the display and additional controls, such as the scroll bar. The additional workload of configuring a scrolling display may make it a less suitable option for ECL systems. (CAP 708)

Problem Statement

Some checklists are lengthy. With a paper checklist, the pilot can look ahead to see how long the list is, and judge how far along he/she is down the list. On an electronic display, the checklist may contain more items than fit on one screen at a time. For task scheduling and other purposes, it is important to know how long the checklist is and how much remains to be completed.

Examples

If the checklist is implemented in terms of discrete "pages," where each page represents the number of items that can be displayed at one time, then the current page and the total number of pages can be indicated using a convention such as "1/3," where the first number is the current page, and the last number is the total number of pages.

If the checklist is implemented on a scrolling window, a side scroll bar can convey all the required information. For example, if half the checklist has been accomplished, the graphical box or bar would be positioned midway down the vertical length of the window.

If the user moves ahead to view later pages or scrolls to a location where the active item is out of view, the active item should not change. If the active item is out of view and the user attempts to change the status of the item, the action should bring the active item into view.

Evaluation Questions
- When viewing looking ahead in a multi-screen checklist, does the active item remain unchanged?
- Is the active item brought into view when it is out of view and the user makes a change to it?

4.3.5 Closing or Completing a Checklist

Equipment Requirements

> - If the ECL system tracks item status and the user attempts to close an incomplete checklist, an indication that the checklist has not been completed should be provided, and the user should be presented with any deferred or incomplete checklist items for review.
> - The user should be allowed to close an incomplete checklist only after the indication has been acknowledged. (See also 2.4.5)

Equipment Recommendations

> - If item status is tracked, the user should receive a positive indication when the checklist as a whole, as well as each item in that checklist, is complete.
> - The action for closing or completing the checklist should be different and clearly distinct from the action of marking an *item* as completed.

Problem Statement

A number of accidents have occurred because of the failure of the flight crew to complete all items on a checklist. Many paper checklists even include "checklist complete" as their last item, to ensure that the checklist is complete. Ensuring that a checklist has been completed is an important error-trapping function that an ECL system should support as well. This includes indicating both the status of the checklist as a whole and each item on the checklist. If a checklist-level indicator is not provided, the crew has to take the time to scan each item in the checklist. If an individual-item indicator is not provided, the crew cannot determine which items are incomplete if no "checklist complete" message occurs.

Examples

If the user attempts to close a checklist that has not been completed, one option is to give remind the user that the checklist is incomplete. The user would then be given the option to either go ahead and finish the checklist or to close it without completion. However, this reminder message may quickly be forgotten by the user.

Another way to tell users that there are pending/incomplete checklists open is to prompt them to view any incomplete checklists when they leave the ECL system (e.g., to view another EFB application), or to highlight incomplete checklists in a view of checklist status indications (e.g., a master list of open checklists).

Reminders to complete deferred or incomplete items could be implemented at a variety of levels. First, there could be visual indicators within the body of the checklist that indicate the presence of deferred items. The item itself could be visually coded as a deferred item or the item could be moved to the end of the checklist. In addition, the checklist itself could have a visual indicator showing that it has deferred items. Also, a code could be used in a master list of open checklists that reflects the presence of deferred items in an open checklist.

Evaluation Questions

- If the user attempts to close an incomplete checklist, are they asked to review deferred and incomplete items?
- Is there a positive indication to the user that all individual items in the checklist are complete, as well as an indication that the checklist as a whole is complete?
- Is the user allowed to close an incomplete checklist after acknowledging that it is not complete?

4.3.6 Closing All Checklists

Equipment Recommendations

> - The ECL should allow a state where there are no open checklists.
> - The system should give a positive indication that no checklists are open; a blank screen is not sufficient, because it could imply a malfunction.

Problem Statement

Checklists are not in use for the majority of a routine flight. With paper checklists, the crew simply puts the checklist back into storage. The act of storing a paper checklist indicates that the checklist has been completed and removes the checklist as a distraction. Removing the checklist can prevent subsequent problems with checklist completion, particularly for sensed items. An equivalent capability should be supported by any ECL system so that the crew knows that no checklists-related tasks are in progress.

Examples

After completing the take-off checklist and passing beyond 18000 feet in US airspace, there are no checklists to use under routine conditions. Paper checklists are then stowed. The ECL system could then display "No currently open checklists" or show an empty master list of open checklists.

The landing checklist may have a sensed items confirming that spoilers are armed and flaps are extended. Shortly after landing, these settings are changed. If the checklist were not closed, it would show the items to be incomplete and incur crew workload by requiring the crews to override the sensed indication.

Evaluation Questions

- Does the ECL allow a state where there are no currently open checklists?
- If no checklists are open, is this state clearly annunciated?

4.4 Interacting with Checklist Items

4.4.1 Indicating the Active Item

Equipment Recommendations

> - The ECL should track and indicate the active item in the checklist in order to help trap potential crew errors.
> - When returning to a checklist that was started but not completed, the item that was active prior to the move should again be active.

Problem Statement

A number of accidents and incidents have occurred at least in part because of the failure of the crew to complete all of the items in a checklist. Distraction, high workload, and other factors may cause the crew to unintentionally skip an item. Similarly, moving between checklists or other EFB functionality can result in the user losing his/her place in the checklist. Even with systems that show individual item status (e.g., completed or deferred), additional time is required to identify the first uncompleted item in the list unless there is the active item is clearly indicated.

Examples

An arrow pointer or outline box could highlight the active checklist item visually.

Evaluation Questions

- Is the active item clearly indicated?
- Does the ECL indicate the item in the checklist where the user was prior to leaving the checklist?

4.4.2 Displaying Item Status

Equipment Requirement

> • If the ECL indicates the status of an item (e.g., active, deferred, overridden, uncompleted, sensed), a clear visual indication of that status should be provided.

Training/Procedures Recommendation

> • In AC 120-64, the term "closed-loop item" is used to describe items in the checklist whose state can be sensed by the ECL system. Note that "closed-loop" does not mean that the ECL system actually *changes* the aircraft state; instead it means that the item status is updated based on the sensed aircraft state. This distinction needs to be clear to the users of the system; the crews need to understand the limits of the ECL automation.

Problem Statement

If the ECL tracks item status, it is useful for the crew to also be shown what the item status is. Because there are potentially several states that an item can be in, it is important that the states are distinguished clearly, otherwise the crew will become confused and may misinterpret the item status, potentially causing distraction or an increase in workload.

Some categories that the ECL may need to distinguish between include active/inactive, deferred, overridden, uncompleted, or sensed. A deferred item is one that the crew has skipped but intends to complete at a later time. An overridden item is one that the crew does not intend to complete. A sensed item means that the ECL system is able to sense either aircraft state or switch position to verify that the item is completed.

Examples

Each checklist item state that is supported by the ECL should have a unique visual code that can be quickly discriminated under a range of lighting conditions. The code that is used should be consistent with other color code applications in the flight deck. Deferred items can also be moved to the end of the checklist to indicate that they remain incomplete.

Evaluation Questions

- How are the possible checklist states indicated and are they easy to recognize?
- Are the visual codes sufficiently unique as to be clearly discriminable under all lighting conditions?
- Are crews aware of the limits of the ECL automation? In particular, are they aware of the limits of the ECL sensing functions?

Electronic Checklist Systems

4.4.3 Moving Between Items Within a Checklist

Equipment Requirements

> - Moving the active-item pointer to the next checklist item should require only a simple action by the user.
> - For ECLs that track the status of individual checklist items, the user should be able to move backward through checklist items to return to a previous item without changing the status of any of the items.

Equipment Recommendations

> - For ECLs that track the status of individual checklist items, the user should be able to:
>
>   (a) Move from an uncompleted checklist item to the next item in the checklist, changing the status of the uncompleted item to "deferred."
>
>   (b) Move to the next item in the checklist automatically after a completing an item.
>
> - The system should allow the user to select one item after another after quickly; system processing should not induce noticeable delays.

Equipment Design Tradeoff(s)

> The system could minimize the potential for skipping beyond the desired item unintentionally by imposing a minimum time lag between cleared items. This may be useful if the input mechanism for marking item completion uses a button that could send a continuous signal, i.e., one that the user can push and hold to send a stream of inputs. By imposing a time lag, inadvertent push and hold entries would not be accepted. However, having a minimum time lag between cleared items could also hinder the crew from completing the checklist as quickly as they need to or could. Instead, the choice of input mechanism could be reconsidered; input mechanisms that allow only one input per actuation may be more appropriate.

Problem Statement

EFB users need to be able to move easily and quickly between items in a checklist. Some users are used to very rapid checklist accomplishment in the paper world and do not tolerate slower response from electronic checklists. Also, if users are not allowed to move around the checklists flexibly, they may take more time to complete the checklist, or they may become frustrated with the ECL system and make more errors using it.

Examples

ECLs that use an active-item pointer should support easy movement to the next item in the checklist. This movement should require only a simple user action. For ECLs that track the status of individual checklist items, moving between items should not affect their status, except in the case of moving to the next checklist item from an uncompleted item. In this case, the status of the prior item should change from uncompleted to deferred.

In addition, the ability to move backward should be implemented to avoid forcing the user to move forward through all checklist items in order to return to the desired item. Backward movement alone should not change a checklist item's status.

On a sensed ECL, a pilot could set an aircraft switch into the correct position, and expect to see the checklist entry updated immediately. If there is a significant delay between the time that the switch is set, and the time that the sensed checklist item changes, pilots may find the checklist performance to be unacceptable.

Evaluation Questions

- Is it easy to move the active-item pointer to the next checklist item?

- Can the user move backward to a previous checklist item without affecting the status of any item? If the user moves forward in the checklist, are deferred items marked appropriately?
- Does the active item change to the next one in the list after an item is completed? Is there a tendency to skip items when attempting to move to the next item?

Electronic Checklist Systems

4.4.4 Specifying Completion of Item

Equipment Requirements

> - The act of specifying that an item has been completed should be simple.
> - When an item is completed, it should not be removed from the screen immediately. The crew should have the chance to review the item, and if necessary, undo their action. (CAP 708)

Equipment Recommendations

> - If active item is indicated by the system:
>
>   (a) When an item has been completed, the next item in the list should become the active item, except if the item is on the next page. A separate action should be required to move to the next page.
>
>   (b) Moving the active pointer to the next item without completing the current item should require an input that is clearly different from specifying that the current item has been completed.
> - Should an item have been incorrectly designated "complete," an easy undo should be available to return the item to either a different status (e.g., deferred), or to the uncompleted status.
> - The completion status of each checklist item should be indicated clearly.

Problem Statement

In order to reduce erroneous entries, crew inputs to ECL systems need to be simple and distinctive (instead of complex and confusable).

An important advantage of electronic checklists is their ability to indicate which items within a checklist have been completed. This feature reduces the likelihood that an item will not be completed. An action separate from that of moving to the next item in the checklist should be required to change an item's status to "completed."

Pilots may make errors while completing a checklist. The ECL should support error recovery by enabling easy modification of an item whose status has been incorrectly modified.

Examples

Changing the status of an item could be implemented by selecting the item to be modified and then selecting the "return to uncompleted status" button. After the item is back in the uncompleted state, the user may change its status again if necessary to get to a different state, e.g., deferred.

The user could press a button to indicate completion of an item. This button should be different from another button that might be used to move to the next item in the checklist without first completing the item (i.e., the other button would mark the task as "deferred" instead of completed).

Evaluation Questions

- Does the action required to change an item's status differ from the action required to move to the next item if the item is not completed?
- Is it easy to change the status of an item to a different status, including uncompleted?
- When the status of an item has been changed to indicate completed, does the next checklist item automatically become active?
- Is a separate action required to move to the next page after all the items on the current page are completed or deferred?
- Is the completion status of each checklist item indicated clearly?

Electronic Checklist Systems 115

4.5 Options

4.5.1 Links Between Checklist Items and Related Information

Equipment Suggestion

> - A set of links to information related to individual checklist items should be provided when that information is available elsewhere in the EFB. The links could direct users to additional information about that item, about the system addressed by the item, and/or to MEL information for that system.

Equipment Recommendations

> - If links to related information are implemented, navigation between the ECL and the related information should be simple and clear to the user. In particular, returning to the checklist item from the related information should be a single-step action.
> - The related information should appear in a single window or area of the screen. If hyperlinks within the related information are activated, the information in that one window (or area) should be updated, rather than opening a separate window with the new content.

Equipment Design Tradeoff(s)

> In choosing the content that is linked to a particular checklist item, designers should think through what type of information is most likely to be needed by the crew. That information should be given easy access. If different types of information will be accessible, the user may need to be able to select the type of information to access.

Problem Statement

Links to related information (e.g., information about that item, the system addressed by the item, and/or the MEL information for that system) can be useful for a variety of reasons. For example, links to MEL information would provide immediate access to the implications of a non-normal condition identified through checklist performance. Less experienced users could also use the supplementary information as a study/review tool.

Examples

A direct link to related information could be implemented by means of a hyperlink initiated from the checklist item. Another way to access related information would be through a pop-up menu that appears over the item when called up. The user could choose the information they wanted to view from this menu.

In either case, returning to the original checklist item should be a single-step action. For example, if a "help screen" is open, there could be a button on it labeled "Return to Checklist". There might also be other links on the help screen, pointing to other related information. If those hyperlinks are selected, no other windows should open; instead, the contents of the help screen should be updated.

If the user is allowed to follow a chain of hyperlinks, it is important that he/she does not become confused as to how to return to the checklist, or to a previously selected link. A standard way of guarding against this type of confusion is to allow only one window to show related information. When a chain of links is followed, the user has the option of returning to the previous link or selecting a new link; either of these actions would replace the content that is currently displayed.

Evaluation Questions
- Does the ECL provide links to useful, related information? Is it easy to select what information to view?
- Can the user return to the checklist from related information in one step?
- Is the related information always shown in one window or area of the screen regardless of how many links were selected?

Electronic Checklist Systems 117

4.5.2 Links to Calculated Values

Equipment Recommendations

> - If the EFB provides calculation worksheets and provides integration features between the application hosting the ECL and the application hosting the calculation worksheets, then
>
>   (a) Direct access to the appropriate worksheet should be provided for all checklist items that can be calculated for both initial calculations, and subsequent review and modification if necessary.
>
>   (b) The user should be able to easily return to the checklist item from which the worksheet was accessed, even if the calculation was not attempted or completed.
>
>   (c) ECL values that were calculated in a linked worksheet should appear in the corresponding checklist location, which should be blank prior to insertion of the calculated value.

Problem Statement

Some checklist items involve setting a system value that has been calculated from one or more tables. The calculation worksheet may be accessible from the corresponding checklist item if the ECL system and the application hosting the calculation worksheets are integrated. These integrations are convenient for the pilot and may reduce workload, but they are not safety enhancements.

Examples

Any checklist field that requires a calculation should be empty until the calculation has been completed. Once complete, the value should appear in all appropriate checklist locations. Access to the calculation worksheet should be supported even after the calculation has been completed to enable the user to review the assumptions on which the calculation is based.

Evaluation Questions

- If the EFB provides worksheets, can the user access and return from these worksheets easily?
- If the EFB supports worksheets that are linked to specific checklist locations, is the checklist item empty prior to the performance of the calculation? Is the checklist item filled in after the performance of the calculation?

4.5.3 Task Reminders

Equipment Suggestion

> - ECLs could provide reminders for tasks that require a delayed action to ensure that the task is completed at the appropriate time. This may be especially useful for non-normal/emergency procedures.

Equipment Recommendations

> - If the ECL system supports task reminders for high priority, time-critical tasks, the reminder should be displayed constantly on the EFB once in progress, and it should attract the pilot's attention at the time that the delayed action should be performed.
>
> NOTE: In order to be displayed constantly on the EFB, the reminder may need to be available even when the ECL is not the currently active EFB application.
>
> - If multiple task reminders can be in progress at one time, crews should be able to determine how many are in progress and to what tasks they refer.

Design Tradeoff(s)

> While task reminders may be useful for specific tasks, they may not be necessary in every case, and should be used judiciously. Otherwise, the crew may be inundated with reminders of competing priorities, increasing workload.

Problem Statement

Pilots may become distracted and forget to complete tasks that require a delayed action (e.g., *stopping* a fuel transfer). ECLs that provide a reminder to complete the delayed action can ensure that the task is completed at the correct time.

Examples

An example of a task that requires time to complete is a fuel transfer. After the fuel transfer is initiated, the crew typically completes other tasks while the transfer progresses, since the transfer may last several minutes. After the correct amount of time has passed, the crew should stop the fuel transfer, but by this time they may be engaged in other tasks. An ECL reminder that attracts their attention at the correct time could ensure that the fuel transfer does not go on too long.

Multiple task reminders could be indicated in several ways. One approach is to use an icon with a number on it that indicates the number of active reminders. The user could click on the icon to access a master list of all active reminders together with pertinent information such as expiration time. Another approach is to use one icon for each reminder. Clicking on the reminder would access more information about that task.

Evaluation Questions

- Does the EFB provide reminders for tasks that require a delayed action?
- Can the user easily review what a reminder is for?
- If task reminders are provided, how are multiple reminders indicated and is it easy to determine what tasks they are associated with?

Electronic Checklist Systems 119

4.5.4 Checklist Branching

Equipment Recommendations

> - When a checklist branches based on a key decision, the selected branch should be clearly indicated.
> - The user should be able to backup to the decision step and choose another decision branch to allow recovery from an erroneous choice.
> - Items that are not on the selected branch should not be selectable.

Problem Statement

While performing a complex checklist, the pilot may have to make key decisions at several points, and choose a branch of that checklist based on each key decision. Keeping track of the active items along a given branch mentally can be cumbersome and error prone.

ECL systems that clearly highlight the selected branch of a checklist can ease the mental burden of keeping track of which items to perform. Also, by clearly encoding the selected branch, the pilot may be more aware of the decision he/she made to select that branch. If items that are not along the selected branch are deactivated, then the pilot cannot mistakenly perform them.

Examples

Decision branching can be indicated by means of a yes/no indicator or graphical means of depicting the alternative choices and checklist branches in response to an explicit, clearly written question. Double negatives should not be used in the question. Based on the user's choice, the user is taken to the next set of items required for that situation. The user should be allowed to change his/her choice by backing up to the decision step in the checklist.

Items that are not along the selected branch could be encoded by a text color, such as a dim gray, or they could be hidden. If they are hidden, the user should be able to view these items for review on request.

Evaluation Questions

- Are decision branches clearly indicated within a checklist?
- Can the user recover from choosing the wrong branch easily?

5 Flight Performance Calculations

Flight performance calculations have been available on the air transport flight deck for several years now. They are often used on portable hardware, e.g., tablet computers, and are now becoming available on installed EFB systems as well. There are relatively few EFB considerations that are *unique* to flight performance applications; these are listed below. In addition, many considerations in Chapter 2 may also apply.

5.1.1 Default Values

Equipment Recommendation

> - If a data entry field is initially blank, then that should indicate that no default value has been assigned by the system.

Training/Procedures Recommendation

> - Crews should be trained to review any default values entered by the system for appropriateness in the current situation.

Equipment Suggestion

> - Default values entered by the system could be highlighted to aid the crew in recognizing them as values that need to be reviewed carefully.

Equipment Design Tradeoff(s)

> Designers should consider carefully what, if any, default parameters are entered for performance calculations. One point of view is that the most commonly used values should appear as defaults, to reduce pilot workload. Another point of view is that default values should represent conservative parameters, so that any results based on incorrect default entries are still acceptable.
>
> EFBs that are connected with other flight deck systems may be able to acquire or verify default values from these systems. If, however, the communication with the other system is lost, designers should have a backup plan for assigning and/or verifying default values.

Problem Statement

Default values can be useful in speeding a routine calculation task. However, default values should be selected carefully because they may not be appropriate for the situation and users may forget to review them. Inappropriate default values could result in significant miscalculations.

Examples

In computing landing performance, the default value for runway conditions could be the one requiring the longest runway length (e.g., poor conditions), or they could be normal conditions (e.g., dry/good).

Evaluation Questions

- Are crews aware that they need to review default values carefully? Can they identify which values are defaults?
- Are any defaults obtained from other flight deck systems? If yes, what is the backup plan for assigning these values if communication with the other system is lost?

Flight Performance Calculations

5.1.2 Data-entry Screening and Error Messages

Equipment Recommendations

- If user-entered data are not of the correct format or type needed by the application, the EFB should not accept the data. An error message should be provided. The message should indicate which entry is suspect and specify what type of input is expected. (AC 120-76A, Section 10.d (3))
- The EFB system and application software should incorporate input error checking that detects input errors at the earliest possible point during entry, rather than on completion of a possibly lengthy invalid entry. (AC 120-76A, Section 10.d (3))
- In order to limit the amount of data that should be re-entered after an erroneous entry is discovered, only that item should be discarded, not the whole set of entries related to the particular task in progress.
- If a required value is missing, the system should produce an error message indicating such, along with the name of the required parameter. The name for the value in the error message should match the label on the input field.

Problem Statement

While pilots who use an EFB may receive some training in use of the device, designers of the system should not expect users to be experts. Well designed error messages help to reduce the training time, promote acceptance of the device, and aid in recovery from errors. Not all entry errors can be caught with data screening, but such screening can be quite effective nonetheless.

Examples

In entering flap settings for a takeoff performance worksheet, values outside the normal range of flap settings (e.g., a three digit, or alphabetic entry) should not be accepted. If an invalid entry is made, the error message could state that a number between 0 and 30 (or whatever the maximum flap setting is for that aircraft) is required.

Evaluation Questions

- Does the system discard entries that are clearly of the incorrect format or type?
- Does the error message clarify the type and range of data expected?
- Are errors in data entry identified at the earlier possible point?

5.1.3 Support Information for Performance Data Entry

Equipment Requirements

> - The units of each variable used in the software should be clearly labeled.

Equipment Recommendations

> - Labels, formats, and units of variables used in the software should match the labels, formats, and units of the data available from other sources of information (e.g., paper reports or flight deck systems). (See also 2.1.3)
> - In performing a task, if the user may need to cross-check with other computations or data stored within the application, the related information should be in view, or else easily accessible.

Equipment Suggestion

> - Definitions of specific terms used in the software should also be easily accessible (e.g., via hyperlinks, or pop-up information areas.)

Equipment Design Tradeoff

> When data entry is required, the EFB designer should consider whether that data will be readily available to the user. If the data are not generally known or easily accessible, the task may increase workload to a level that is unacceptable.

Problem Statement

Units used for data (e.g., degrees Fahrenheit or Centigrade) must always be visible in order to prevent confusion and ensure accuracy of calculations. Labels, formats, and units used in calculations should be consistent with other sources of information, again, to prevent confusion.

In addition, the user should not have to manipulate what information is in view in order to complete a single logical task. System designers should consider whether the data requested will be readily available to the user, and whether the format available to the user matches the format expected by the software. If the data are not readily available, the user may make errors, or may not catch errors made previously. If the necessary data are difficult to access, speed and accuracy of task performance will suffer, and pilot workload will increase.

Examples

It may be possible to use different data properties (e.g., format) to distinguish between measures in different units. For example, temperature could be entered as 12F to indicate a Fahrenheit value, and 12C to indicate a Centigrade value.

Different airline operators may use different terminology to refer to variables used in various computations, such as weight and balance. The terminology used in the software should match that used by the airline in any other paperwork (e.g., maintenance, dispatch release forms, or fuel records).

In computing various flight performance data, the user may need to refer to route information, which should be in view or readily accessible. Also, it may be useful to have definitions for terms such as "damp" runway available, especially if those terms affect the calculations.

Evaluation Questions

- Are units clearly labeled?
- Does the terminology used in the software match the terminology used in other operator documents?
- Is all the information necessary for a given task presented together, or easily accessible?

5.1.4 When and How to Do Performance Calculations

Training/Procedures Recommendation

> - In order to reduce the possibility of errors, users should learn to complete performance calculations at particular times within the flight. They should also plan to complete the calculations in a certain order, and through a certain set of ordered steps.

Equipment and Training/Procedures Recommendation

> - In designing the flow of steps for completing a task in the software, designers should consider whether that flow is logical to the user and can be trained easily. Similarly, training on how to do a performance calculation should be matched with the steps in the software.

Equipment Suggestion

> - The software could prompt or remind the user of the order of steps for completing a task.

Problem Statement

Performance calculations are usually completed in a series of steps. In some cases, the steps should be completed in a particular order, but in other cases, the order of steps is not critical. When the order of steps is important, the software should prompt the user for information in that order. Even if the order of steps is not critical, users should be trained on the process for completing the task so that they are less likely to forget individual steps.

Users should also be trained on when to initiate and complete performance calculations with respect to the flight timeline.

The software should be designed to match the user's expectations with regard to the order of steps in the process, and the training should be designed to reinforce this procedure for completing a calculation.

Examples

Weight and balance computations are usually done at the gate while passengers are boarding. However, the weight and balance data may change significantly at the last minute, and so should be easily modifiable just prior to takeoff.

Evaluation Questions

- Are users trained on when and how to do performance calculations?

5.1.5 Modifying Performance Calculations

Equipment Recommendations

> - The software should allow the user to make modifications to previously computed results quickly.
> - Once the user begins modifying data that were used in earlier calculations, the earlier output should be erased to make clear that those values are no longer valid.

Problem Statement

In practice, flight performance calculations often need to be updated at the last minute, e.g., during taxi out, which can create a high workload situation. Therefore, the performance calculation application should allow the user to make modifications easily. For example, the user should be able to modify one or two parameters only, then have the calculations redone, without re-entering all the data.

Examples

Modifications due to weather changes are fairly common. For example, rain may begin and a dry runway may become wet during taxi out. Rapid wind shifts may also occur, for example, if a gust front passes through.

Evaluation Questions

- Can the user modify performance calculations easily?
- Are outdated results of performance calculations erased when modifications are entered?

Flight Performance Calculations 127

5.1.6 Use of Performance Calculation Output

Equipment and Training/Procedures Recommendation

> • Results computed by the flight performance application should be displayed in a manner that is understood easily and accurately.

Training/Procedures Recommendation

> • End users should be aware of any assumptions upon which the flight performance calculations are based.

Problem Statement

Flight performance software may be more or less complex, depending on the end user. For general aviation operations, a simple weight and balance module can be quite useful. For air transport, a more full-featured application is necessary. Simpler applications may need only some basic aircraft information and it may provide output that is relatively simple to understand. More complex applications may need more detailed information about the aircraft, and it may provide output that is much more detailed.

In all cases, users should be aware of any assumptions the software is making, and they should be able to interpret system output easily and accurately.

Examples

The weight and balance module of a general aviation flight performance calculator needs to know specific information about that aircraft (e.g., empty dry weight). The user should ensure that this information is accurate for their situation.

For Part 121 operations, the flight performance calculations need to take into account NOTAM and MEL restrictions. For example, users should be able to select MEL items and the application should apply appropriate performance penalties. If a runway is shortened via NOTAM, the user should also be able to enter that information into the application.

If the application computes V-speeds for an FMS-equipped aircraft, the output should list the V-speeds in the same order as the input page of the FMS.

Evaluation Questions

- Are the results of the flight performance calculations understood easily and accurately?
- Are users aware of any assumptions on which the calculations are based?

5.1.7 Coordination Between Flight Crews and Ground Dispatch Personnel

Training/Procedures Recommendation

> - Procedures should be in place to define and coordinate any new roles the flight crew and dispatch may have in creating, reviewing and using performance calculations supported by EFBs. (AC 120-76A, Section 10.f (4))

Problem Statement

In many airline operations today, ground dispatch personnel are responsible for completing flight performance calculations. The flight crew reviews and approves the flight plan that the ground dispatcher has prepared. These roles may be altered by the introduction of an EFB application that allows flight crews to perform the calculations. In order to prevent any confusion about who is responsible for completing the calculations, or reviewing them, the new roles of the flight crew and dispatcher should be coordinated. Both the flight crew and dispatcher should be working from a common set of data, and they should both have a common flight plan prior to the flight.

Examples

The flight crew may be able to use an EFB to modify or check a flight plan quickly just prior to departure, taking into account any last minute changes to the weight of the aircraft, or any last minute changes in the departure runway (e.g., to check if they can take a shorter runway). If dispatch is not involved in checking these last minute changes, they may need to be informed by the crew at their earliest convenience as to the changes that occurred. The crew may inform dispatch via voice, or possibly through data link.

Evaluation Questions

- Are the roles of dispatchers and flight crews coordinated?

5.1.8 Aircraft Performance Documentation

Training/Procedures Requirement

> - Procedures should be developed to ensure that any information required to be available outside of the aircraft is transferred from the EFB at the appropriate time and place. (See, for example, 14 CFR Part 121.697)

Equipment Suggestion

> - Even if there is no legal requirement to be able to transfer data off of the EFB, the ability to do so is desirable.

Problem Statement

EFBs may be used to compute flight performance data. In some cases (Part 121 operations), copies of this computed information are required to be deposited at a ground station prior to takeoff. Therefore, there should be a procedure for transferring the data from the EFB to some other media (e.g., paper), or onto a different ground computer.

Part 91 operators may not be legally required to transfer data off the EFB, but would benefit from the ability to do so.

Examples

Weight and balance data should be available in either paper or electronic format to airline personnel other than the crew. One way of handling this requirement is to send the information electronically from the EFB to company flight dispatchers. An alternative would be for the EFB to print out the required information onto paper, which would then be left at the point of departure.

Evaluation Questions

- What is the procedure for ensuring that, if necessary, EFB data can be stored outside of the device?

6 Electronic Charts

Eventually, sophisticated electronic charts may replace paper charts in the flight deck. However, in this report it is assumed that there will be a significant transition period over the next few years where both paper and electronic charts co-exist on the flight deck. (Even if the paper charts are removed from the flight deck, most pilots are so familiar with using paper charts that it will take some time for them to become as comfortable with electronic charts as they are with paper charts.) This guidance covers the transition period between paper and electronic charts more so than it covers the end-state that many hope for, where integrated electronic charts are the norm. The discussion here is limited to stand-alone terminal charts, i.e., charts that are used for operations in the terminal area, such as approach charts, airport taxi charts, and arrival/departure procedures. The guidance does not address en route charts or integration issues (e.g., integration of text updates such as and NOTAMs and operator-tailored information or integration of live data, such as traffic, weather, or aircraft state). Also, the guidance does not address charts with display of ownship, which are considered to be moving map displays. Guidance for moving map displays is provided in RTCA DO-257A, which is titled "Minimum Operational Performance Standards for the Depiction of Navigational Information on Electronic Maps." Another source for information on electronic chart display issues is an Aerospace Recommended Practice (ARP) document being drafted by the SAE G-10 Electronic Chart committee. The draft ARP presents guidance on information priorities for database driven electronic charts, which will be especially useful for systems that are aware of aircraft state, and could use that information to declutter the charts automatically.

To provide a framework and context for these considerations, background on aeronautical charts, both paper and electronic, is presented in Section 6.1 below. General human factors considerations for "first generation" electronic charts are presented in Section 6.2. While guidance for approval of electronic charts is provided in this chapter, evaluators and designers should keep in mind that any aeronautical charts used in high workload situations, such as takeoff and landing operations, will require significant evaluation (see 2.1.2 Using EFBs During High Workload Phases of Flight).

6.1 Background

6.1.1 Paper Charts

Aeronautical charts are a rich source of information for conducting flight operations. They provide a variety of information that, in many cases, is required for flight (see 14 CFR Parts. 91.103 Preflight Action; 121.549 Flying Equipment; 125.215 Operating Information Required and 135.83 Operating Information Required). They are standard equipment in any pilot's flight bag.

Pilots use aeronautical charts as both a planning tool and as a source of specific, detailed reference data (Wright and Barlow, 1995). As planning tools, charts help the pilot to envision the procedure to be flown by providing information such as which navigation aids will be used, and/or altitudes and headings to follow for a published approach and landing procedure. Charts also contain a great deal of reference data such as minimum parameters for landing under different weather and equipment conditions, terrain elevations, and radio frequencies.

Different types of charts are used in different stages of the flight and under different operating environments. Under visual flight regulations (VFR), color VFR sectional charts are used. Under instrument flight regulations (IFR), approach plates provide specific and detailed information for each published approach to a runway end. IFR en route charts provide long range navigation information for IFR operations. These typically place less emphasis on visual navigation than VFR sectional charts and therefore provide less detailed terrain information. In IFR operations, pilots may also have to follow published arrival and departure procedures, which depict how the aircraft is expected to enter the traffic pattern at busy airports.

To date, charts are generally used in paper form. Paper aeronautical charts have evolved into their current size and configuration. They present a large amount of information in a relatively small space, using special symbols and drawing conventions. The symbols and drawing conventions sometimes vary between chart manufacturers, so pilots typically use charts from a single manufacturer for most, if not all, of their flights. While there are many similarities among charts produced by different manufacturers, there can be small, but significant, variations. Many nation states, including the United States government, publish charts in accordance with standards set by the International Civil Aviation Organization (ICAO). Independent chart providers, such as Jeppesen Inc., are not required to follow the ICAO standards, but generally do so.

From an operator's point of view, the main disadvantages of paper charts include (a) their weight and bulk, and (b) the time, manual effort, and resulting cost and potential for human error associated with keeping paper charts up to date. Paper charts often weigh several pounds by themselves, particularly if the aircraft is used for global operations. Manual updating of paper charts is a tedious process and charts could easily be updated incorrectly.

6.1.2 Types of Electronic Charts

Recently, there has been significant industry development of electronic charts for EFBs. A lightweight EFB could store and display charts while taking up less volume and weight than the equivalent set of paper charts. In addition, electronic charts are desirable because they can be updated electronically, and because they could support new, dynamic functionalities, such as real-time displays of own aircraft position along the route of flight, if approved for use. Another advantage of electronic charts is that data on the chart could be de-cluttered for a given operation. De-cluttering could be performed in advance for non-interactive charts, or it could be performed interactively by the user if the chart is based on an underlying database of elements. If the EFB is aware of aircraft status, automation could drive the chart de-cluttering in addition to manual inputs.

Here, we distinguish between three electronic charts formats: *raster, vector, database*. *Raster* charts are created by scanning a paper chart (or portions thereof) into a bitmap format. If the full chart is scanned, the whole chart will be non-interactive. However, if portions of the chart are scanned separately, there could be limited interactive (non-paper-like) features, such as user control over what chart portions are displayed (e.g., as in JeppView FliteDeck 1.0). In this case, portions of the display (e.g., the plan view) are stored as objects, but the objects themselves are non-interactive like paper. On vector charts, every symbol on the display is defined by a latitude and longitude and stored as an individual object. As a result, lines and symbols can be redrawn from vector descriptions as the user zooms in and out of the chart, which could produce a clearer image when the chart is zoomed to different degrees. In a *database-driven* electronic chart individual display elements are stored in the database. These charts can support a great deal of user interactivity. For example, the user could filter out obstructions that are below his/her current altitude. Other, even more flexible de-cluttering options could also be supported.

On the horizon are electronic charts with automation. If the EFB is able to sense the aircraft state (e.g., position and intended flight plan), automation could become especially powerful. The automation could include algorithms that dynamically configure the display for the aircraft's actual position. For example, the system could remove all information for an arrival procedure that is not in use. As automated electronic charts are implemented, it will be important to understand their overall impact in the flight deck. These automated charts could support a new range of capabilities, but the logic used would need to be designed, tested, and evaluated carefully.

Another potential advantage of these automated electronic charts is that charts may not be discrete documents any more, but instead may become part of a complete map, combining approach charts, en route charts, and so forth into a seamless whole. This integration may change the way pilots access needed information, and may enable them to look beyond the boundaries of a given discrete chart to see how it relates to other charts and features of the area.

6.1.3 Comparison of Paper and Electronic Charts

A major advantage of paper charts is the operational utility and usability has been established through years of in-flight use and evolution. In contrast, there is little documented operational experience (e.g., regarding learning curves) with electronic charts. In particular, there is a need for more data collection and analysis of how often different chart data and chart types are used in order to design good user interfaces for electronic charts (see 2.4.1 User Interface Design-General). For example, in what order are charts usually used? What information is "most important" and when? How do usage patterns vary based on type of operation (e.g., Pt 91 versus Pt 121/135 operations, use of charts at familiar versus unfamiliar airports, or even use of charts during short-haul versus long-haul flights).

Electronic Charts 133

Paper and electronic charts are compared and contrasted along several dimensions in Table 6-1 and Table 6-2. Notice that several issues for electronic charts are raised. These issues are discussed later as well, in the associated considerations.

Updating	**Paper Charts**
	Manual updates. Prone to human error. Time-consuming
	Electronic Charts
	Integrity of data should be checked prior to distribution. Automated updates. Prone to automation errors. Quick. (See 2.4.15, 2.4.16, and 2.4.17 for related guidance.)
Clutter	**Paper Charts**
	Clutter is unavoidable because several procedures are commonly incorporated on one plate to minimize the volume of paper. In some cases, clutter can be controlled through good design and high level of customization for a specific audience, but not always. Complexity of content will affect how much information is displayed, indirectly affecting clutter.
	Users search through data (e.g., landing minima) that may be irrelevant to their specific operation.
	Clutter cannot be controlled by user.
	Electronic Charts
	Potentially less clutter because may be easier to customize for specific audiences (e.g., only show data applicable to a particular operation) automatically.
	User may have some control over clutter with a database-driven chart (e.g., select whether some elements are shown or not). However, there is additional workload associated with configuring the display to bring up desired information. Some of the workload of configuring the electronic chart could be offset by an intelligent chart system that correctly anticipated user needs for information.
	Charts that store information in raster or vector format, instead of in a database format, would not be able to support a flexible, user-controlled de-cluttering capability.
Legibility of Individual Symbols/ Characters	**Paper Charts**
	Symbology on paper charts is designed for legibility at relatively close viewing distances (approximately 12-15 inches). Reference data is usually smaller in size than procedural information.
	Symbology also optimized for use on standard size chart (approximately 5 inches by 7 inches).
	The user can position the paper chart at a comfortable distance to optimize legibility.
	Electronic Charts
	May be difficult to read small characters/symbols if they are transferred directly (i.e., the same size, stroke width, etc., are retained) from paper charts. Paper highlighting techniques (e.g., bolding) may not transfer well to lower resolution electronic displays.
	If the EFB cannot be repositioned physically, the smallest characters/symbols may need to be larger than they are on paper charts.
	The resolution of the electronic display and other characteristics of the display technology will affect symbol/character legibility (see 2.4.10 through 2.4.13 and 2.5.3).
	Characters/symbols could be modified for better legibility on electronic displays with significantly lower resolution than paper.

Table 6-1 A comparison of paper and electronic chart features (continued on next page).

Use of Individual Charts	Paper Charts	
	User are highly familiar with paper charts (may be especially familiar with one particular manufacturer's charts).	
	Electronic Charts	
	May require training to learn how to use individual charts, depending upon similarity to paper charts.	
	New functions could be supported (e.g., integrate with real-time display of own aircraft position); these may also require some training.	
Accessing Charts	Paper Charts	
	Familiar process for accessing individual charts. The process relies on the human knowing what chart is necessary.	
	Charts expected to be used in a particular flight can be set aside in advance (e.g., destination arrival procedure).	
	The time to bring up a desired chart can be lengthy because of the large number of paper charts that have to be searched.	
	Electronic Charts	
	May require training to learn how to access a desired chart.	
	Potentially quicker access (on the order of a few seconds), but poor implementations (e.g., lots of steps) could increase workload.	
	Potentially new ways of accessing individual charts (e.g., custom searches for charts matching specific criteria).	
	Also, if connected with FMS, could potentially assist in error checking by double-checking user's selection and asking for confirmation of selection if inconsistency is found.	
Accessibility of Information	Paper Charts	
	Both planning and reference information is available all in one place. No need to reconfigure the display to access any particular type of information.	
	Information is constantly available; no need to "share" the display with other functions.	
	Electronic Charts	
	May require training on how to configure display and how to bring up desired reference data quickly	
	May need to obscure the chart to access other EFB functions.	

Table 6-2 A comparison of paper and electronic chart features (continued from previous page).

6.2 General

6.2.1 Transition from Paper to Electronic Charts

Equipment Recommendations

> - The information structure of electronic charts should match that of paper charts, particularly if paper backups are still in use. That is, information grouped in paper charts should also be grouped in electronic charts.
> - The visual structure of electronic charts, including symbology and general layout, should be compatible with that of paper charts, although it is not necessary to copy the visual structure of paper charts exactly.

Equipment Design Tradeoff(s)

> Some visual elements of paper charts (e.g., methods of highlighting information, fine line widths, or font choices and sizes) may not transfer well to lower resolution electronic displays directly. Modifications may be necessary to maintain legibility of the electronic chart.

Training/Procedures Design Tradeoff(s)

> Pilots may need training to learn how to configure and use individual electronic charts, depending upon their information and visual similarity to paper charts. Any functions supported on electronic charts that are not available on paper charts may also require pilot training.

Equipment and Training/Procedures Design Tradeoff(s)

> If paper backup charts are not in use, the electronic chart may be modified significantly from the familiar format. Before the new chart format is adopted, the design should be studied thoroughly in terms of its usability. Issues such as training time, learning curves, potential negative transfer from extensive familiarity with use of paper charts, and overall task performance should be considered and evaluated.

Problem Statement

Pilots are highly familiar with the information and visual structure of paper charts. These users have developed highly efficient and individualized strategies for retrieving chart information for reference and planning purposes. These strategies are so well ingrained that pilots can have difficulty switching between paper charts from different sources, which may vary relatively little in format. Similar to the transition that they have to do when switching from, for example, US government charts to Jeppesen charts (or vice versa), users will need to spend time developing and learning new strategies for using electronic charts. If the electronic chart is created based on a totally new structure, developing these strategies may be challenging at first, and the challenges may last for a long time. Also, confusion and errors are more likely if pilots do not find the electronic information where they expect it to be based on their experience with paper charts. In some cases, there may even be negative transfer from paper charts, meaning that habits that helped pilots to use paper charts actually hurt their performance with electronic charts if the formats are contradictory.

There can be many features on electronic charts that are not possible on paper. Such electronic features can add value, but the fundamental chart structure should be maintained. The more consistent the information and visual structure of electronic and paper charts are, the easier it should be for pilots to transition to the new charts and take advantage of the new functions without increasing workload and errors.

Examples

Paper charts are divided into sections (e.g., plan view, profile view, minimums table, etc.). These sections should be available as units in electronic charts as well, although they may not be visually arranged in the same way as they are on a paper chart.

In order to be compatible with the visual structure of paper charts, which are usually oriented vertically (i.e., in portrait mode), electronic charts could also be presented in portrait mode.

A symbol on a paper chart may have more detail than the corresponding electronic symbol, but the basic shape should be recognizable across the two media.

Pilots are used to seeing paper charts in north-up orientation. If electronic charts are track up, their orientation should be explicitly indicated so that pilots do not revert to their traditional expectations based on paper charts.

If the pilot is used to looking in one place for a particular item (e.g., the top of the page) and that item is shown on a different part of the electronic chart, errors, higher workload, and delayed task completion may occur. A well thought out framework for the new layout on the electronic charts could alleviate these potential problems.

Evaluation Questions
- Do the electronic charts match paper charts in terms of their visual and information structure?
- How difficult will the transition to electronic charts be for pilots who are highly familiar with using paper charts?

6.2.2 Updates to Electronic Charts

Equipment Recommendations

> - Corrections and updates to electronic charts should be made within the electronic chart application, unless they are of a temporary nature.
> - Corrections or updates that contain high priority or time-sensitive data should not be made via paper notifications to flight crews. (See Sections 2.4.16 Updating EFB Data and 2.4.17 Crew Confirmation of EFB Software/Database Approval for more information)

Problem Statement

In order to rely on electronic charts, pilots should be sure that, as long as the database is current, the charts are accurate and up to date. Pilots should not have to remember to check for more recent paper corrections or updates to charts. Paper amendments to electronic charts could lead to confusion and errors, particularly in high workload non-normal conditions.

Examples

Changes in the final approach course or runway length may need to be updated on the charts. If the changes are permanent (at least relative to the update cycle of the charts), these changes should be made through an electronic update.

Evaluation Questions

- Are corrections and updates to electronic charts made electronically? If not, how are updates handled?

6.2.3   Hard Copy Backups of Electronic Charts

Equipment Recommendations

> - If hard copies are used as backups for electronic charts, the quality of the hard copy should be of sufficient quality that it could be used as effectively as the original paper chart. In particular,
>
>   (a) The hard copy should be legible; all chart details should be clearly visible.
>
>   (b) The quality of the paper should be acceptable for normal use.
>
>   (c) If the electronic charts use color, but only a monochrome hard copy can be generated, all the information should be preserved and distinguishable in the monochrome version.
>
>   (d) All of the chart information should be able to fit on one printed page.
>
>   (e) The physical hard copy should be at least as large as a standard paper chart (5" x 7" for a US government chart), and could be larger for better legibility.
>
>   (f) Selection of the output size, if more than one option is available, should be made by the user.

Problem Statement

AC 120-76A mentions that during the transition to a paperless flight deck, an operator will need to establish a reliable means of providing the information required by regulations to the flight crew (Section 9). As a backup for paper charts, a flight deck printer or fax machine could generate hard copies of all applicable data required for the flight. However, flight deck printers and fax machines are relatively expensive and uncommon technology to date. More common flight-deck printing technology supports only text, not graphic output, which is not useful for generating backup charts. If graphics-capable flight-deck printers are used, however, it is important that their output can be used as effectively as a paper chart, or else the output is not a satisfactory substitute.

Examples

If chart details are not clear on the hard copy, the hard copy will not be usable in flight because pilot workload and head-down time will increase, and errors are more likely.

In order to obtain satisfactory quality of printouts, the charts should be printed out at a size that is equal to or larger than its original paper equivalent. Unless the information on the chart is substantially de-cluttered, charts that are printed at a reduced size are likely to be more difficult to read and use than current paper charts are.

It may not be necessary to store an entire set of backup paper charts in the flight deck. Instead, only charts that are expected to be used for that flight (e.g., for the origin, destination and alternate airports) could be printed out from an EFB onto a flight deck printer a short while before the flight.

Evaluation Questions

- Are charts printed from an electronic chart application of sufficient quality that they are as usable as the original paper documents?

Electronic Charts 139

6.2.4 Scale Information

Equipment Requirements

- Chart scale information should always be visible for charts that are drawn to scale.
- Chart scale information should be accurate at all times. When the display is zoomed, chart scale information will need to be updated to stay accurate.
- Static chart scale information that may come as part of the paper chart database should be removed from the electronic display unless it will always be accurate (i.e., the display cannot be zoomed and it is drawn at the correct scale).
- If the chart is drawn "not to scale" (e.g., an arrival or departure procedure) a label indicating that fact should always be in view.

Problem Statement

Pilots are familiar with the scales used on paper charts. For example, on a U.S. government VFR sectional chart, 1 inch represents 7 NM. They can use this information to estimate distances. If zooming is implemented, however, 1 inch may represent some distance other than the standard shown in the scale. Because pilots should would be misled by inaccurate scale information and because they may make incorrect assumptions if the scale information is not shown, the chart scale should be visible and accurate at all times. Note that scale information could be presented in a variety of ways, e.g., a range ring, or a line segment whose length represents a standard distance.

Examples

A pilot may use a variable zoom in to view a detailed area of the chart to gather specific reference information, such as the frequency for a navigation aid. At this zoom level, which is not a standard preset, a physical distance of 1" on the screen may represent an unusual actual distance, such as 3.8 nautical miles. Instead of depicting the scale as 1" = 3.8 NM, one option is to show a line that is the length needed to represent a typical distance, such as 5 NM, and leave out the text description.

Scale information is also provided in the chart from the latitude marks. These marks stay conformal when the display is zoomed, so they are not misleading.

On arrival and departure procedures a "Not to Scale" label appears as part of the chart database. However, if panning is implemented, this label can go out of view, which it should not. The "Not to Scale" label should always be in view.

Evaluation Questions

- Is accurate chart scale information always visible?
- Is potentially inaccurate static scale information removed from the display?

6.2.5 Basic Zooming and Panning

Equipment Recommendations

- If the chart application supports zooming, it should also support panning. Similarly, if the chart application supports panning, it should also support zooming.
- If zooming and panning are implemented, the chart's visual edges should be clearly marked. Visual edges should only be shown when there is no further information outside that area so that the user knows that absence of an edge indicates the existence of off-screen information.
- Panning should be designed such that the user always knows which way to move to bring more of the chart into view.
- The user should not be allowed to pan to an area so that no portion of the chart will be displayed.
- If the zoom level can be configured by the user, the user should be able to return to default settings easily.
- If the display can be panned, the user should be able to return to a central default view easily.
- Zooming and panning operations should not result in lengthy processing delays. (See Section 2.4.6, Responsiveness)

Equipment Suggestions

- Zooming should be implemented in concert with panning such that the user can view a region of interest and return to the larger chart view easily. This is especially important on small displays where it is more difficult to maintain one's orientation when zoomed in tightly.

Equipment Design Tradeoff(s)

Zooming in helps the user to see detail and zooming out helps the user to get context. However, the task of setting the zoom level can induce workload. The design of the zoom mechanism and controls can affect the how much workload is induced.

Designers also need to consider the controls that are implemented for controlling zooming and panning. The choice of controls, their directionality, and their labels can all influence how intuitive they are for the majority of users.

One possible consequence of being zoomed in is that the pilot could lose awareness of proximal, but off-screen threats such as terrain or obstructions. If the electronic chart application can sense aircraft state, i.e., ownship position, the system could potentially generate alert the pilot to off-screen proximal threats, but this is a fairly complex function that will need to be designed carefully.

Problem Statement

Electronic displays do not yet have a resolution that is close to that of paper. In order to clearly show all the detail in a paper chart, the EFB display would have to be much larger than is feasible for a typical flight deck. Given the display size and resolution constraints, zooming and panning are the only realistic way of ensuring that all the chart information is accessible. Zooming in increases display legibility both because the display elements become larger in size, and because there are usually fewer elements in the area of interest, producing a less cluttered display. Zooming out allows the user to see where he/she is in the context of the full chart, but at the expense of seeing all the detail. While zooming is a useful feature, it needs to be designed both with performance and workload issues in mind.

Examples

Pre-defined zoom levels that the user can select from may generate less workload than variable zoom features, but they also limit the customizability of the display, which could reduce its usability. Magnifying-glass-style zooming is useful in that it allows the user to see detail while retaining context, and is fairly intuitive to use.

Evaluation Questions
- Does the zoom feature allow the pilot to view a specific region of interest easily?
- Can the user orient him/herself easily using zooming and panning together?

6.2.6 Procedures for Use of Electronic Charts in Part 121/135 Operations

Training/Procedures Recommendations

> - For multi-function EFBs, Part 121/135 operator's policy should specifically define what other applications can or cannot be open and in use while electronic charts are in active use (i.e., during taxi, takeoff, or on approach and landing).
> - The operator's policy should also address any special procedures that may apply to use of charts with multi-function EFBs that have knowledge of the intended flight plan.
> - The operator's policy should also specify under what conditions charts that are not actively in use can be viewed for any EFB (not just multi-function EFBs).

Problem Statement

As stated in Section 2.3.1, Part 121/135 operators need to have a statement of their EFB operating policy. Because electronic charts are a high priority source of information during certain phases of flight, the policy statement should specifically address the priorities of the different EFB applications relative to charts. Even within the charting application, the operator's policy should specify under what conditions charts that are not actively in use can be viewed.

Examples

Flight crews may need to perform flight calculations during an approach. The operator's policy should specify whether the approach plate can be out of view while using the flight performance application. An operator may choose to resolve this by stating that two EFB display units are required in the flight deck, and that one should always display the chart in use while the other EFB accesses any other applications.

Some EFBs may have knowledge of the intended flight plan through a data connection with other flight systems. This would allow the next chart to be pre-selected for manual call-up at the appropriate time or the chart could be called up automatically. These procedures should be well defined.

Evaluation Questions

- Does the Part 121/135 operator's EFB policy specifically call out what other applications can be used when the electronic charts are actively in use?
- Does the operator's policy address any special procedures that apply to the use of electronic chart applications that are aware of aircraft state?
- Does the operator's policy specify under what conditions charts that are not actively in use can be viewed?

Electronic Charts 143

6.2.7 Orientation of Electronic Charts

Equipment Requirements

- Orientation of the charts (track/heading-up or north-up) should be indicated continuously, unless the visual format differences in the two orientations are so salient that pilots can easily and reliably recognize which orientation is in use.
- When charts are oriented with respect to directionality (e.g., track or heading), and directionality information becomes unusable, it should be clear to the pilot that that information is not available. (See also 2.4.9 on Display of System Status).

Equipment Recommendations

- When charts are oriented with respect to directionality (e.g., track or heading), and directionality information becomes unusable, the application should:

  (a) Notify the crew of the unusable directionality data (i.e., give a positive indication of the failure). The error message should also inform the crew that as a result of the failure, the charts must revert to north-up orientation.

  (b) After the failure notification is acknowledged by crew input, the chart orientation should revert to north-up, the indication of chart orientation should be updated to "north-up", and any cues that could imply that directionality information is available should be removed. (See also 6.2.10)

  NOTE: Further guidance from DO-257A may be applicable (e.g., Sections 2.3.1.2 and 2.4.1.2).

- All text and symbols (e.g., navaid or waypoint symbols) other than those designed to reflect a particular compass orientation should remain upright at all times to facilitate readability. (DO-257A)
- Crew input should be required to change the orientation of the electronic charts (e.g., between track-up and north-up), to ensure that the crew is aware of the change.

Installation Recommendation

- If the EFB supports a track/heading-up mode, it should be installed such that the top of the EFB display is aligned with the aircraft longitudinal axis to eliminate an extra mental rotation step and to reduce the potential for misleading the pilot.

Equipment Design Tradeoff(s)

Creating electronic track/heading-up charts is more complex than creating electronic north-up charts because the designers need to consider how to handle rotation of all the display elements separately.

Equipment and Training/Procedures Design Tradeoff(s)

Electronic charts could be presented north-up only, or give pilots the option of showing north-up or track/heading up. Since paper terminal charts are drawn north-up, using this type of chart orientation should be familiar to pilots. However, it is not clear how pilots will use track/heading-up electronic charts. In particular, these orientations could potentially induce new types of errors, particularly on aircraft with other displays of spatial information (e.g., displays of traffic and/or weather, PFDs, or EGPWS). (See also 6.2.8.)

Equipment and Training/Procedures Design Tradeoff(s) (continued)

> Since north-up paper charts are the norm, track/heading-up electronic charts are really enhancements over paper functionality. As such, there may be new training and procedural issues as well as software design issues if track/heading-up modes are supported.

Problem Statement

Unlike paper charts, whose orientation is static (and usually north-up), the orientation of an electronic chart could change between north-up, track-up, or even heading-up. One or another of these orientations may be more suitable for a given situation. For example, navigational moving maps are generally shown with track or heading up while north-up is usually reserved for a planning mode where the pilot is looking many miles ahead of his/her current location. Track/heading-up displays are usually better for making tactical (near-term) decisions, such as which way to turn at the next waypoint along the intended route. Given that map orientation may be variable, the electronic chart application should clearly indicate the current orientation at the minimum. If the electronic chart application supports both north up and track/heading-up orientations, and the two modes are confusable, the pilot could end up following an incorrect path leading to a variety of potentially severe outcomes. Variable map orientations also introduce other types of design issues and complexities, as noted above.

Examples

It may be useful to allow a temporary track-up selection that reverts to north-up when the pilot releases the control. This would allow pilots to adjust the display orientation for making specific comparisons without introducing the potential complexity of orientation modes.

If users can choose between north-up and track/heading-up orientations for electronic charts, they may need some guidance as to when each of these modes is appropriate.

Evaluation Questions

- If the EFB display supports more than one orientation, is the orientation in use evident from the display behavior, and/or a mode annunciation?
- Could the pilot become confused about the display orientation and make errors? What types of errors could result from incorrect interpretations of the display orientation?

6.2.8 Using Charts with Other Flight Deck Displays of Spatial Information

Equipment and Training/Procedures Recommendation

> - In order to ease comparisons between spatial displays (e.g., navigation, traffic, and weather displays), there are three key factors to consider: map scale, map orientation, and database quality. Pilots should be aware of how flight deck displays of spatial information vary in terms of these parameters at all times in order to integrate the different sources of information into a coherent picture.

Problem Statement

Pilots may need to compare, correlate, and mentally consolidate information from EFB electronic charts with information from other flight deck displays of spatial information, such as navigation, traffic, weather, or terrain displays. These comparisons are often made mentally. However, incompatibilities between these displays in terms of their scale, orientation, and even their underlying databases could make integration of information challenging and even error prone. Integrated displays of all these types of spatial information could be the eventual solution for this problem, but these are typically complex and expensive.

Examples

The pilot may use a traffic display to identify and follow another aircraft into an approach procedure. This may require the pilot to integrate information between the EFB electronic chart and the traffic display. In particular, map scale and orientation must be taken into account accurately in order to understand where the other traffic is in relation to the approach procedure that own aircraft is following.

If some information is duplicated on spatial displays in the flight deck (e.g., terrain is shown both on the EFB electronic chart, and on EGPWS), pilots could be confused by the fact that it is possible for the displays to provide different information (in terms of resolution or accuracy) if the underlying databases come from different sources.

If the EFB is connected to the FMS, it could run in a "slaved" mode where the EFB electronic chart is automatically matched with the FMS flight plan. This would ease comparisons between the electronic chart and navigation display. In addition to this automated functionality, however, pilots may also need to retain the ability to select an electronic chart manually in order to access charts for reference purposes.

Evaluation Questions

- Is the pilot always aware of the differences in map scale, orientation, and database quality between the various displays of spatial information on the flight deck?

6.2.9 Access to Individual Charts

Equipment Requirements

- The chart label of the currently selected chart should be displayed continuously.

Equipment Recommendations

- The system should allow users rapid access to charts that are pre-selected by the user.
- The chart application should promote good error management, meaning that it should help the crew ensure that the correct chart was selected and, if an error was made, it should allow common corrections to be made quickly.
- The chart application should support multiple search methods (e.g., by name, by geographical region, by present position).
- Search results should be ordered intuitively, with the most likely selections at the top of the list, and least likely selections at the end of the list. Even if the most-likely selections are unknown, unlikely selections that are known should be at the end of the list. Electronic chart applications that have knowledge of the flight plan could order search results even more intuitively.
- Since last minute runway changes are a high workload event, the electronic chart interface should make selection of alternate runways as easy as possible during the approach.

Training/Procedures Recommendation

- Flight crews may need training to learn how to access a desired chart because there are potentially many ways of accessing individual charts (e.g., custom searches for charts matching specific criteria), and some of these methods are new to electronic charts.

Equipment Suggestion

- If available, connectivity with the FMS can make it easier to access the right chart because the system would know useful information, such as the origin and destination airports, and the route of flight. The pilot would not have to make duplicate entries to get this information into the EFB.
- If there is connectivity with the FMS, and the crew has to divert while en route, selecting a diversion airport in the FMS could cause the EFB electronic charts to pre-select the approach charts associated with that airport for manual call-up.
- If aircraft position is known, the distance to various airports could be computed, allowing the electronic chart application to show the "nearest" airport in case an emergency landing is necessary. Similarly, if the aircraft is far from its departure field and would not be returning there even in an emergency, charts for that airport could be removed from the pre-selected list.

Electronic Charts 147

Equipment and Training/Procedures Design Tradeoffs(s)

> Pilots could potentially access with electronic charts quicker than paper charts, but poor implementations for accessing electronic charts (e.g., too many steps) could increase overall workload.
>
> Ideally, charts should be available according to normal usage patterns for that operator, if those use patterns can be established in advance. Part 121/135 may have more well-established usage patterns than Part 91. The individual operator's preferences could even be implemented in the software.
>
> Accessing approach charts in a new way (e.g., using an electronic search function instead of paper) may make new kinds of errors possible. The potential for errors made possible by new functions such as the search and access methods should be examined.

Problem Statement

Pilots need to be able to access charts quickly and accurately. Putting charts into an electronic medium makes new, potentially better, methods for accessing charts possible. Among these is the ability to search for charts and access them in multiple ways, rather than just based on their relationship with a given airport. Access could also be intelligently assisted by the system.

Examples

The electronic chart application could suggest charts to the pilot as an ordered list of choices based on current position, destination, frequency of use, and likelihood of use based on the current flight plan.

If the EFB is aware of the final destination, it could provide a list of approaches for that destination.

Since last minute runway changes are a high workload event, it might be handy for the electronic chart application to make selection of alternate runways as easy as possible during the approach. One way to do this would be for the alternate approaches to an airport to be available on tabs or in a selection box so the pilot can see the options and select the desired one with one step, rather than having to reselect the airport to get a list of the alternate arrivals.

It may be possible to search for a departure or arrival by some means other than through the particular airport that the departure or arrival is attached to. For example, the pilot may want to search by name of the arrival procedure, without specifying the airport.

Evaluation Questions

- Can pilots select charts quickly and accurately?
- How can users identify errors in chart selection? Is it easy to make common corrections?
- Are there multiple ways of accessing charts?
- If FMS connectivity, or sensing of aircraft state is used to facilitate access to charts, does the application anticipate user needs appropriately, while allowing the user the flexibility to deviate from system suggestions easily?

6.2.10 Knowledge and Display of Own-Aircraft Position

Equipment Requirements

> NOTE: If ownship is presented on a map display, requirements from TSO-C165 and RTCA DO-257A apply and should be met.
>
> - Display of ownship position should not be supported on non-georeferenced or not-to-scale terminal charts (e.g., arrival and departure procedures).

Equipment Recommendations

> - The range of display zoom levels should be compatible with the position accuracy of the ownship symbol.
> - If the chart is zoomed or panned such that ownship symbol is no longer in view, there should be a clear indication of the direction in which own aircraft lies.

Training/Procedures Recommendations

> - Pilots should be aware of the use, accuracy, and limitations of the display of own aircraft position.
> - Pilots may not have much experience using north-up moving map displays. There may be need for a familiarization period with these displays. Training needs should be assessed for the specific situation.

Equipment Design Tradeoffs(s)

> Display of own aircraft position on a chart can serve as a crosscheck for the pilot's internal assessment of his/her position, but pilots could become over-reliant on the display, promoting a false sense of security. Therefore, the immediate display of any system failure in this capability becomes especially critical (see 2.4.9).
>
> If the electronic chart application has knowledge of own aircraft position, it could use this information to configure the chart automatically, even if own aircraft position is not displayed. However, it is not clear what would be the best way to use this information.
>
> On a north-up moving map display, designers must consider how to keep the ownship symbol in view; the map may need to shift as the ownship goes beyond the displayed area.
>
> Different issues may be encountered when designing the vertical profile portion of an approach plate as part of an electronic chart, particularly if own aircraft position is displayed. See RTCA DO-257A for guidance on these issues.
>
> The size and shape of the ownship symbol should be chosen with care because it could potentially imply something about the precision of its displayed position.

Problem Statement

Accurate display of own aircraft position can be extremely useful to the pilot in crosschecking his/her internal estimate of ownship position. However, it is important that the pilot understand when own aircraft position may be in error or unavailable so that he/she does not become overly reliant on this information.

Georeferenced charts are drawn to scale accurately so that own-aircraft position can be correlated with a particular location on the chart. However, some charts, such as those for arrival and departure, are not usually drawn to scale. Own aircraft position should not be displayed on charts that are not drawn to scale, or those that are not georeferenced because the information would be misleading, conveying the sense of greater accuracy than exists.

Electronic Charts   149

Examples

   Training and operational procedures should emphasize that the primary flight display is to be used for navigation, not display of own-aircraft position on electronic charts.

   If the update of own aircraft position is lost, the own aircraft symbol could be removed from the display, and an explanatory message could appear on the EFB.

   If the location of a ground track depiction on a chart is accurate within several hundred feet, it should not be possible to zoom in so close that accuracy within a few feet is implied.

   An ownship symbol in the shape of a cross hair may imply a high degree of precision, where a circular shape may not.

   One method for indicating the loss of directionality information may consist of changing the ownship depiction from a directional symbol to a non-directional symbol (e.g., circle). (From DO-257A)

   Equipment that does not have access to heading information may derive track based on changes in position over time (e.g., a Global Navigation Satellite System (GNSS) sensor used to derive track). However, this information will become unreliable when the taxi speed is low relative to turning velocity. (From DO-257A)

   The zoom feature should not allow the pilot to reduce the scale beyond the point where the display of ownship position is accurate.

Evaluation Questions

- Is display of ownship only supported on georeferenced charts?
- When own aircraft position is displayed, is the displayed position accurate to within the scale of the chart?
- If the pilot zooms the chart display to show a close in view, does the display of own aircraft position remain accurate?
- Are pilots aware of the limitations of the display of own aircraft position?

6.2.11 De-cluttering and Display Configuration

Equipment Recommendations

> - If there is a de-clutter capability, it should not be possible for the pilot to remove safety critical display elements (e.g., terrain, obstructions, or special use airspace) without knowing that they are suppressed. If such information can be de-cluttered, it should not be possible for the pilot to believe that it is not visible because it is not there.
> - Managing the display configuration (e.g., scale, orientation, and other options and settings) should not induce significant levels of workload. That is, routine display configuration changes should be minimized. (See also 2.1.1.)
> - The information prioritization scheme should be documented in the pilots' guide and in the certification plan.

Equipment Suggestion

> - If there is a zoom feature, there should also be a coordinated de-clutter feature so that the display remains usable when it is zoomed out (i.e., shows a large area). If there is no de-cluttering when a large area is in view, there may be so many small objects on the display that none of the information is useful.

Equipment and Training/Procedures Design Tradeoffs(s)

> Pilots may need training on how to configure and de-clutter the display. More training will be necessary for more complex user interfaces.
>
> Manual control over the display of individual information elements (e.g., obstructions, or navigational aids) could become complex. On the other hand, automatic display configuration could be frustrating to use, and potentially unsafe, if it does not match the user's expectations.

Problem Statement

Display decluttering is likely to be necessary on EFB electronic charts, given the typical display sizes and resolutions available. This feature needs to be implemented appropriately to ensure that errors and pilot workload are not induced. Decluttering is one example of a display configuration task. Other display-configuration tasks that pilots may have to perform include manipulating the display orientation, zoom level (scale), or any other custom settings, e.g., color codes.

Examples

One display-configuration settings may be to show only navigation aids and the approach path, with terrain and obstruction data removed. There should be a clear indication that safety-related data are missing from the display. Ideally, the whole format of the display should appear distinctly different from the usual format, rather than the indication being limited to a small message or symbol in a corner of the display.

Many new navigation display formats allow pilots to selectively display "layers" containing different types of objects: navaids, terrain, airways, etc. It is possible that some electronic chart implementations will allow the same type of control over what is displayed and what is not displayed. If so, the most important layers for the intended use of that display should not be allowed to go out of view.

Evaluation Questions
- Is there a clear indication when safety-related display elements are suppressed?
- Does configuration of the display, including decluttering induce significant workload?

Electronic Charts 151

6.2.12 Color Coding of Chart Symbols

See 2.4.3 for general information on the use of color.

Equipment Design Tradeoffs(s)

> Paper charts may make extensive use of color. However, it may not be appropriate to transfer all of the paper chart colors directly to the electronic display. The colors may not be as usable on the electronic display because (a) they may not be as discriminable as they are on paper, and (b) the range of color available for printing inks is wider than that for electronic displays.
>
> Individual display elements may eventually be colored. However, no standards have been developed to date on what colors should be used, or how display elements should be grouped in terms of color.
>
> If both specific symbol and background elements (e.g., terrain) are color coded, layering systems will need to be developed to ensure that every element appears in its correct color.

Problem Statement

Color could be used for a variety of purposes on a terminal chart, such as displaying terrain elevations or for denoting information about specific display elements. Because of the large amount of information on a chart, it may be tempting to use color in many ways, especially since color is being used in paper charts. In the electronic display, however, using too many colors will make the display more difficult to use, not less. Instead of trying to match all the color used on some paper charts, it is important that use of color in the electronic chart application be consistent with other color usage on the EFB. In particular, as noted in 2.4.3, use of red and amber should convey warning and caution information only.

Examples

Brown is preferred for terrain (over green) because green has the connotation of "okay". Brown is consistent with some paper charts, and also with the attitude display. However, green is used to depict safe terrain in EGPWS.

Airports may be color coded based on weather conditions or utility for the current flight. If color codes are used, a redundant text code could be used, such as the addition of a text string to the airport identifier (e.g., BOS-VFR).

Evaluation Questions

- Are red and amber used appropriately in electronic charts?
- Are electronic chart display elements color-coded? If so, is the electronic chart color code consistent with other EFB color usage?

7 References

Chapter 1: Introduction

Chandra D. C. & Mangold S. J. (2000). *Human factors considerations in the design and evaluation of electronic flight bags (EFBs) Version 1: Basic functions.* (Report No. DOT-VNTSC-FAA-00-22). Cambridge, MA: USDOT Volpe Center.

FAA AC 120-76A. (July 2002). Guidelines for the certification, airworthiness, and operational approval of electronic flight bag computing devices.

FAA Policy Statement ANM-99-2, Guidance for reviewing certification plans to address human factors for certification of transport airplane flight decks.

FAA Policy Statement ANM-01-03, Factors to consider when reviewing an applicant's proposed human factors methods for compliance for flight deck certification

FAA AC 25-11 (16 July 1987) Transport Category Airplane Electronic Display Systems.

FAA AC 23-1311-1A (12 March 1999) Installation of Electronic Displays in Part 23 Airplanes

FAA AC 91.21-1 (20 August 1993) Use of Portable Electronic Devices Aboard Aircraft

Chapter 2: EFB Systems

General

FAA AC 120-53. (May 13, 1991) Crew Qualification and Pilot Type Rating Requirements for Transport Category Aircraft Operated Under 14 CFR Part 121.

FAA AC 25.1523 (2 February 1993) Minimum Flight Crew.

MILSTD1472F (23 August 1999) Department of Defense Design Criteria Standard Human Engineering.

Ahlstrom, V. and K. Longo (Eds.) Federal Aviation Administration Technical Center (May 2003). *Human Factors Design Standard for Acquisition of Commercial-off-the-shelf Subsystems, Non-Developmental Items, and Developmental Systems.* (Report No. DOT/FAA/CT-03/05, HF-STD-001) Available at http://hf.tc.faa.gov/hfds/

General Aviation Manufacturers Association. (October 2000) Recommended practices and guidelines for Part 23 cockpit/flight deck design. http://www.gama.aero/home.php

SAE (2003) *Human Factors Criteria for the Design of Multifunction Display for Civil Aircraft*, ARP 5364.

SAE (1988) *Human Engineering Considerations in the Application of Color to Electronic Aircraft Displays*, ARP 4032.

SAE (1991) *Electronic Display Symbology for EADI/PFD*, ARP 4102/7.

SAE (1996) *Human Engineering Recommendations for Data Link Systems*, ARP 4791A.

Crane, J.M., Bang, E.S, and Hartel, M.C. (1994). *Standardizing interactive display functions on the 777 flight deck.* In Proceedings of Aerotech 1994. (Paper No. 942093) Society of Automotive Engineers: Warrendale, PA.

Cardosi, K. and D. Hannon (1999) *Guidelines for the use of color in ATC displays.* (Report No. DOT/FAA/AR-99/52). Cambridge, MA: USDOT Volpe Center.

System Messages, Errors and Alerts

FAA AC 20-140 (16 August 1999) Guidelines for Design Approval of Aircraft Data Communications Systems.

FAA Terrain Awareness and Warning System (TAWS) Technical Standard Order TSO-C151b http://av-info.faa.gov/tso/

SAE (1988) *Flight Deck Alerting Systems*, ARP 4102/4

References

Graphical User Interface Style Guides

Microsoft Corporation. *The Windows™ Interface Guidelines for Software Design.* (1995). Redmond, Washington: Microsoft Press.

Apple Computer, Inc. *Macintosh™ Human Interface Guidelines.* (1992). Reading, Massachusetts: Addison-Wesley Publishing Company.

Open Software Foundation (1993). OSF/MOTIF™ Style Guide (Revision 1.2), Englewood Cliffs, NJ: Prentice Hall.

Chapter 3: Electronic Documents

NASA/FAA Operating Documents Project (October 2000). Developing Operating Documents: A Manual of Guidelines. http://human-factors.arc.nasa.gov/opdoc-workshopV/relateddocs.htm

Dillon, A. (1994) *Designing Usable Electronic Text.* Taylor & Francis: Bristol, PA.

Rosenfeld, L. & Morville, P. (1998) *Information Architecture for the World Wide Web.* O'Reilly & Associates: Sebastopol, CA.

McKinley, T. (1997) *From Paper to Web.* Adobe Press, Macmillan Computer Publishing: Indianapolis, IN, USA

Chapter 4: Electronic Checklists

Civil Aviation Authority (2000, December), Guidance on the design, presentation and use of electronic checklists. (CAP 708). [Available through Documedia, 37 Windsor Street, Cheltenham, Gloucestershire GL52 2DG, UK telephone 44-1242-235151, fax 44-1242-584139]

Civil Aviation Authority (2000, December), The development of guidance on the design, presentation and use of electronic checklists. (CAA Paper 2000/9). [Available through Documedia, 37 Windsor Street, Cheltenham, Gloucestershire GL52 2DG, UK telephone 44-1242-235151, fax 44-1242-584139]

Boorman, D. (2001) Today's electronic checklists reduce likelihood of crew errors and help prevent mishaps. ICAO Journal, Vol. 56, No. 1, pp. 17–21.

Boorman, D. (2000) Reducing flight crew errors and minimizing new error modes with Electronic Checklists. Proceedings of HCI Aero 2000, Toulouse, France.

FAA AC-120-64 (April 24, 1996) Operational Use and Modification of Electronic Checklists.

FAA Office of Safety Services (1996) Human performance considerations in the use and design of aircraft checklists.

Degani, A. (1992). *On the typography of flight-deck documentation.* (NASA Contractor Report 177605). Moffett Field, CA: NASA Ames Research Center.

Degani, A. & Weiner, E.L. (1990). *Human Factors of Flight-Deck Checklists: The Normal Checklist.* (NASA Contractor Report 177549). National Aeronautics and Space Administration Ames Research Center: Moffett Field, California.

Chapter 6: Electronic Charts

Wright, M. W. and Barlow, T. (March 1995) *Resource Document for the Design of Electronic Instrument Approach Procedure Displays.* DOT/FAA/RD-95/2. Available through http://www.ntis.gov/.

RTCA/DO-257A (June 2003) Minimum Operational Performance Standards for the Depiction of Navigation Information on Electronic Maps

FAA Draft Technical Standard Order. Electronic Map Display Equipment for Graphical Depiction of Aircraft Position TSO-C165 http://av-info.faa.gov/tso/Tsocur/C165.doc

# Appendix A:
EFB Industry Snapshot

The EFB market is growing and diverging into specialties (e.g., hardware, software, integrated solutions, content). More companies are developing EFBs, and more customers are showing interest in purchasing EFBs. The purpose of this industry review is to provide a primer on who is involved in the industry and what their efforts are. This informal summary of EFB technology provides a picture of the current state of EFB development as of September 2003, and reflects the large growth in the industry in the past two years. In the future, companies may change their strategies and areas of specialization. If available, updates to this industry review will be posted at the Volpe EFB website (www.volpe.dot.gov/opsad/efb/).

This industry snapshot consists of three parts:

1. Industry Review: This review provides information about EFB systems, software/content, and hardware that are currently on the market or in active development. This material was gathered through industry contacts, demonstrations, websites, brochures, and trade journal reports. Attempts were made to verify the information with a company representative, but this was not always possible. Also, since manufacturers change their systems continuously, the accuracy of the information in the industry review cannot be guaranteed. Links to product websites are embedded within the text and provided in a table at the end of this section, but these may also become out of date fairly quickly.

2. Military Activities: The US Department of Defense is showing interest in EFB technology. A military conference on EFB was held at the US Air Force Academy (USAFA) in March 2003. Highlights from the conference and an overview of efforts in the military domain are discussed.

3. Industry References: The references provided here consist of trade journals and military publications where more information about industry and military activities can be found.

1 Industry Review

This review provides detailed information about EFB systems, software/content, and hardware. This review was prepared in September 2003. As with any system development cycle, changes in the design occur frequently; as a result, accuracy of the information cannot be guaranteed. For each product, the manufacturer's website is provided where more recent information can be found.

We begin with three tables that provide an overview of the manufacturers included in the industry review. Table 1 lists EFB system manufacturers, Table 2 lists software and content providers, and Table 3 lists hardware manufacturers. In each table, the products developed are classified as a function of their target market: air transport, business jet, high end GA, low end GA, or military. The remainder of this review consists of three sections.

1.1 EFB systems. For each system, the review describes the display characteristics, controls, mounting style, applications supported, approvals obtained, potential customers, websites where more information can be found, and an image of the product. Some airlines are working with system integrators on higher end functionality (e.g., communicating electronic information on/off the aircraft, allowing the integration of aircraft operations with existing legacy systems for maintenance, operations, planning, etc.). This functionality is described in AC 120-76A but is not covered here. Panel-mounted systems with EFB-like functions (e.g., Innovative Solutions & Support Integrated Cockpit) are also excluded from this review.

1.2 Software and Content Providers. For some of the systems described in Section 1.1, additional applications can be purchased separately from software and content providers. The software options available are presented in Section 1.2. A table listing the software and content applications available for use on EFB platforms is provided. The table lists information on the application type, hardware and software compatibility, and potential customers.

1.3 Display Providers There are several display companies who have developed portable computing devices that have been used as EFB platforms. The software described in Section 1.2 may be installed on these devices in order to develop a complete EFB system. These hardware options are described in Section 1.3. Information provided for each display includes display characteristic, user interface options, mounting style, approval (if any), potential customers, websites where more information can be found, and an image of the product.

Table 1. Overview of EFB Systems

MANUFACTURER	Air Transport	Business Jet	MARKET General Aviation		Military
			High End	Low End	
ADR					
• FlightGuide(FG) 3600		✓	✓	✓	✓
• FG-5000		✓	✓	✓	✓
• FG-8000	✓	✓	✓	✓	✓
ApproachView TD-840			✓	✓	
Astronautics Pilot Information Display (PID)	✓				✓
Avrotec					
• FMP 200			✓	✓	
• FMP 300		✓	✓		
Boeing / Jeppesen / Astronautics	✓				✓
C-Map/Aviation AvMap					
• EKP-II/NT-Pro			✓	✓	
• EKP-III C/III C Pro			✓	✓	
CMC Electronics (formerly Northstar)		✓	✓		✓
e**flight**systems, LLC			✓	✓	
Flight Deck Resources	✓	✓	✓	✓	✓
• SkyTab 770					
• SkyTab 800					
• SkyTab 900R					
GSCS					✓
JP Instruments			✓	✓	
NavAero	✓	✓	✓	✓	✓
Paperless Cockpit	✓	✓	✓	✓	✓
• LT P-600					
• GA-EFB 4000					
Teledyne Technologies (formerly Spirent Systems)	✓	✓	✓		✓
Universal Avionics					
• Universal Cockpit Display	✓				
• Universal Cockpit Display Terminal (UCDT-II)		✓			✓

Table 2. Overview of Software and Content Providers

MANUFACTURER	MARKET				
	Air Transport	Business Jet	General Aviation		Military
			High End	Low End	
Adobe	✓	✓	✓	✓	✓
Airbus	✓				
Aircraft Data Fusion	✓				
Aircraft Management Technologies (AMT)	✓	✓	✓		✓
Astronautics	✓	✓			✓
ControlVision				✓	
Echo Flight		✓	✓	✓	
Flight Deck Resources		✓	✓	✓	
Flytimer				✓	
Honeywell	✓	✓	✓	✓	✓
ION Systems	✓	✓	✓	✓	✓
Jeppesen	✓	✓	✓	✓	✓
LIDO	✓				
On Board Data Systems	✓	✓	✓	✓	
RMS Technology		✓	✓	✓	
Rockwell Collins	✓				
Stenbock & Everson				✓	
Teledyne (formerly Spirent Systems)	✓	✓			✓
WSI			✓	✓	
WxWorx			✓	✓	

Table 3. Overview of Display Providers

MANUFACTURER	Air Transport	Business Jet	MARKET General Aviation		Military
			High End	Low End	
Astronautics	✓	✓			✓
CMC Electronics (formerly Northstar)		✓	✓		✓
Fujitsu	✓	✓	✓		
HP/Compaq laptops, IPAQ	✓	✓	✓	✓	
Panasonic	✓	✓	✓	✓	✓
Paperless Cockpit	✓	✓	✓		✓
Teledyne Technologies (formerly Spirent Systems)	✓	✓			✓
WalkAbout Computers	✓				

Appendix A   159

1.1 EFB Systems

This section provides a list of EFB systems in development or in use. The information provided for each system includes:

- Product name
- Website(s) where more information can be found. The text in the following table is hyperlinked to the manufacturer's site. A list of URLs is included at the end of this appendix.
- Display size and resolution
- Controls, i.e., how the user interacts with the device
- Mounting style
- Applications supported
- Approvals
- Potential customer(s)

1. ADR

FlightGuide-5000

FlightGuide-8000

Photos courtesy of ADR.

Product Name(s)	FlightGuide (FG)-3600, FG-5000, FG-8000
Website(s)	<ul><li>ADR</li><li>FG-3600</li><li>FG-5000</li><li>FG-8000</li></ul>
Display Size	FG-3600, FG-5000: 8.4", 800x600 SVGA backlit and polarized active matrix FG-8000: 8.4", 800x600 SVGA backlit and polarized active matrix
Controls	Vertical touch screen; FG-8000 also has mouse control
Mounting Style	Cockpit-mountable; FG-8000 also yoke-mountable
Form Factor	Fujitsu tablet computer
Operating System	Microsoft Windows
Applications Supported	Open architecture – any Windows-compatible software
Electronic Charts	✓ (JeppView FliteDeck)
Electronic Checklists	✓
Electronic Documents	✓
Flight Performance Calculations	✓
Flight Planning	✓ (JeppView FliteStar and FliteMap)
Video Surveillance	
Other	Moving maps and GPS, WSI real-time satellite weather, terrain avoidance, Internet and satcom communications, scheduling programs
Approvals	Hardware Class 1 or 2 EFB
Potential Customer(s)	Air transport, business jet, high-end GA. Jet Aviation installed ADR EFB in Boeing Business Jets

2. ApproachView

Photos courtesy of ApproachView.

Product Name	ApproachView TD-840
Website(s)	• ApproachView • Fujitsu Stylistic LT P-600 pen computer
Display Size	8.4" diagonal display; 800x600
Controls	Touch screen
Mounting Style	R-A-M ball. See www.ram-mount.com for options.
Form Factor	Full screen, tablet-size pen computer (Fujitsu Stylistic LT P-600)
Operating System	Microsoft Windows 2000
Applications Supported	
Electronic Charts	✓ (Jeppesen FliteDeck)
Electronic Checklists	
Electronic Documents	
Flight Performance Calculations	
Flight Planning	✓ (Jeppesen JeppView, FliteMap)
Video Surveillance	
Other	
Approvals	
Potential Customer(s)	High- and low-end GA

Appendix A   161

3. Astronautics

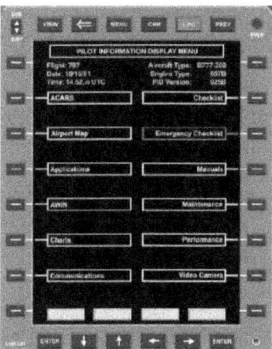

Photo courtesy of Astronautics.

Product Name	Pilot Information Display (PID)
Website(s)	<ul><li>Astronautics</li><li>EFB effort: PID</li><li>July 2002 Avionics Magazine article</li><li>August 2003 Avionics Magazine article</li><li>Named Product of the Year by Flight International</li></ul>
Display Size	8.4" or 10.4"
Controls	16 soft keys, 12 dedicated bezel keys, brightness increase/decrease keys, touch screen, and external keyboard
Mounting Style	Fixed or adjustable. One display for each pilot, adjustable arm, window mount, or tethered
Operating System	Linux and Microsoft Windows 2000
Applications Supported	Open architecture
Electronic Charts	✓ (e.g., Jeppesen EFB)
Electronic Checklists	✓
Electronic Documents	✓ (e.g., Jeppesen EFB)
Flight Performance Calculations	✓ (e.g., Jeppesen EFB)
Flight Planning	✓
Video Surveillance	✓
Other	Aviation weather (AWIN), data link (SATCOM, GateLink etc.), CPDLC, CDTI, & Taxi Position Awareness (moving map)
Approvals	Available as Class 2 or Class 3 EFB<ul><li>Certified (Tailored Linux) and Uncertified (Microsoft Windows) applications for Class 3 PID</li><li>Certified (Linux) or Uncertified (Microsoft Windows) applications for Class 2 PID</li></ul>TSO Certified Hardware Level C, D and E software supported (DO-178B). Certification completed.
Potential Customer(s)	Air transport and Business Jet. 767 and Challenger systems scheduled to be operational in January 2004

4. Avrotec

FMP 200

FMP 300

Photos courtesy of Avrotec.

Product Name	FMP 200, FMP 300
Website(s)	• Avrotec • FMP 200 • FMP 300
Display Size	10.4" diagonal
Controls	FMP 200: Touch pad and keyboard FMP 300: Pilot Input Pointing System
Mounting Style	Panel
Operating System	
Applications Supported	
Electronic Charts	✓
Electronic Checklists	
Electronic Documents	
Flight Performance Calculations	
Flight Planning	✓
Video Surveillance	
Other	FMP 200: Echo Flight: EchoMap Data Link Weather Information FMP 300: Compatible with StormScope
Approvals	FMP 300: TSO for C110A and C113
Potential Customer(s)	FMP 200: low-end GA FMP 300: business jet and high-end GA

Appendix A 163

5. Boeing, Jeppesen, and Astronautics

Boeing, Jeppesen, and Astronautics have put together a cooperative effort to develop an EFB. In their agreement:
- Boeing will do the aircraft integration
- Jeppesen will implement the software, which may also marketed to other EFB platforms
- Astronautics will implement the hardware

Photo courtesy of Jeppesen, Inc.

Product Name	EFB
Website(s)	<ul><li>Boeing<ul><li>Information for Boeing customers (password and account required)</li><li>Boeing EFB effort</li></ul></li><li>Jeppesen</li><li>Astronautics<ul><li>Astronautics EFB effort</li></ul></li><li>July 2002 Avionics Magazine article</li></ul>
Display Size	10.7" diagonal display for a panel mount version in 777, 767, 757, 747, and 737 aircraft
Controls	16 soft keys, 12 dedicated bezel keys, brightness increase/decrease keys, touch screen, and external keyboard
Mounting Style	Installed – one display for each pilot
Operating System	Linux or Microsoft Windows 2000
Applications Supported	Open architecture
Electronic Charts	✓
Electronic Checklists	✓
Electronic Documents	✓
Flight Performance Calculations	✓
Flight Planning	✓
Video Surveillance	✓
Other	aviation weather (AWIN), data link (SATCOM, Connexion by Boeing, GateLink etc.), terrain, CDTI, Logbook Fault Finder, and Taxi Position Awareness (moving map application)
Approvals	Hardware Class 3: Certified (Tailored Linux) and Uncertified (Microsoft Windows) applications DO-160 tested Level C, D and E software supported (DO-178B). Certification completed
Potential Customer(s)	Air transport: KLM and Pakistan Airlines expect system to be operational in October, 2003, on their new Boeing 777s

6. C-Map/Aviation

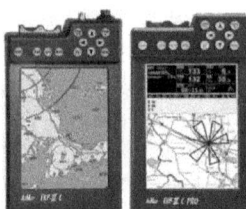

EKP III-C   EKP-IIIC Pro
(for GA)   (for professionals)

Images courtesy of C-MAP/Aviation

Product Name	AvMap EKP-IIIC/EKP-III C Pro
Website(s)	• C-Map/Aviation
Display Size	640x480
Controls	12 dedicated buttons
Mounting Style	Kneeboard
Operating System	C-Map Proprietary
Applications Supported	
Electronic Charts	✓
Electronic Checklists	✓
Electronic Documents	
Flight Performance Calculations	✓
Flight Planning	✓ (AvMap Flight Planner)
Video Surveillance	
Other	Moving map, ownship position, HSI display
Approvals	
Potential Customer(s)	High- and low-end GA; law enforcement professionals

7. CMC Electronics (formerly Northstar)

Photo courtesy of CMC Electronics Inc.

Note: CMC is currently working on the next generation of the CT-1000.

Product Name	CT-1000
Website(s)	• CMC Electronics (formerly Northstar) • CT-1000
Display Size	6.4" diagonal, 800x600 resolution, 256 colors
Controls	12 soft function keys; dedicated keys allow for zoom, dim, bright, and other functions; mouse and pen
Mounting Style	Mounted in aircraft but removable
Operating System	Microsoft Windows OS
Applications Supported	
Electronic Charts	✓
Electronic Checklists	✓
Electronic Documents	
Flight Performance Calculations	✓
Flight Planning	
Video Surveillance	
Other	moving map
Approvals	DO-160 tested
Potential Customer(s)	Air transport, business jet, helicopters for paramilitary missions

Appendix A   165

8. e**flight**systems, LLC

Photos courtesy of e**flight**systems, LLC.

Product Name	eflightpad
Website(s)	• eflightsystems
Display Size	8.4" TFT color LCD, 800x600
Controls	Touch screen, numeric keypad, four cursor buttons
Mounting Style	
Operating System	Microsoft Windows OS
Applications Supported	
Electronic Charts	✓ (Jeppesen FliteDeck)
Electronic Checklists	
Electronic Documents	
Flight Performance Calculations	
Flight Planning	✓ (Jeppesen FliteMap)
Video Surveillance	
Other	
Approvals	
Potential Customer(s)	High- and low-end GA

9. Flight Deck Resources

SkyTab 770

SkyTab 800

SkyTab 900R

Photos courtesy of Flight Deck Resources.

Product Name	SkyTab 770, SkyTab 800, SkyTab 900R
Website(s)	• Flight Deck Resources • SkyTab 770, SkyTab 800, SkyTab 900R
Display Size	SkyTab 770: 8.4" XGA TFT Color SkyTab 800 & 900R: 10.4" XGA TFT
Controls	Touch screen: All SkyTab models Onscreen keyboard: All SkyTab models Full keyboard: SkyTab 900R
Mounting Style	Certified and non-certified mounts/attachments
Form Factor	Full screen tablet PCs (Proprietary SkyTab 770 has no moving parts, SkyTab 800 uses Fujitsu ST 4120P and SkyTab 900R uses Panasonic Toughbook 18)
Operating System	Microsoft Windows XP Professional
Applications Supported	
Electronic Charts	✓ (Chartrax™ and JeppView with FliteDeck)
Electronic Checklists	✓
Electronic Documents	✓
Flight Performance Calculations	Aircraft Performance Group
Flight Planning	✓
Video Surveillance	✓
Other	NEXRad weather: WSI and WxWorx™
Approvals	FAA-AEG Approved
Potential Customer(s)	Major and regional airlines, other certificated carriers, corporate aviation, and low-end general aviation.

Appendix A

10. GSCS

SAMM LTP 600

SAMM MAV

SAMM Remote

Photos courtesy of GSCS.

Product Name	SAMM family
Website(s)	• GSCS • SAMM
Display Size	8.4" diagonal display
Controls	touch screen
Mounting Style	
Form Factor	SAMM LTP 600: Fujitsu LT P 600 (full screen, tablet size, pen computer) or kneeboard computer SAMM MAV: Xybernaut MAV PC (wearable, voice activation, head mounted display), or kneeboard PC SAMM Remote: Remote display for LTP 600, MAV or other PC
Operating System	Microsoft Windows operating system
Applications Supported	
Electronic Charts	✓
Electronic Checklists	✓
Electronic Documents	✓
Flight Performance Calculations	✓
Flight Planning	✓
Video Surveillance	✓
Other	Moving map
Approvals	Flight clearance
Potential Customer(s)	Military, homeland security, commercial

11. JP Instruments

Image courtesy of JP Instruments website.

Product Name	Nav 2000
Website(s)	• JP Instruments • NAV 2000
Display Size	6.5" or 10" diagonal tethered or self-contained unit, 640x480 VGA
Controls	Touch screen, mouse, remote keyboard
Mounting Style	Vehicle mounts available
Operating System	Microsoft Windows OS
Applications Supported	
Electronic Charts	
Electronic Checklists	
Electronic Documents	
Flight Performance Calculations	
Flight Planning	
Video Surveillance	
Other	Moving map software including FliteMap and RMS Vista
Approvals	
Potential Customer(s)	High- and low-end GA

Appendix A   167

12. NavAero

tPad 800

t-Bag Ultra

t-Bag Executive

t-Bag Premier

Photos courtesy of NavAero.

Product Name	t-Pad 800, t-Bag Ultra, t-Bag Executive, t-Bag Premier
Website(s)	• NavAero
Display Size	t-Pad: 8.4" diagonal, 800x600 color LCD t-Bag: 8.4" diagonal, 800x600, or 10.4" diagonal, 1024x768
Controls	Touch screen display
Mounting Style	Kneeboard or certified mount
Operating System	Microsoft Windows
Applications Supported	Any Microsoft Windows compatible software
Electronic Charts	✓
Electronic Checklists	
Electronic Documents	✓
Flight Performance Calculations	✓
Flight Planning	✓
Video Surveillance	✓
Other	ARINC, moving map, weather, terrain avoidance, scheduling
Approvals	Class 2 EFB certification of navAero t-Bag EFB family is in process.
Potential Customer(s)	tPad 800 display currently in-use in high end and low end GA aircraft tBag Ultra: high-end and low-end GA tBag Executive: business jet tBag Premier: air transport and military

13. Paperless Cockpit

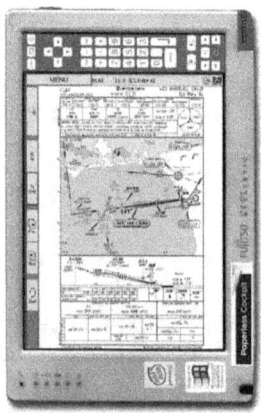

Paperless Cockpit LT P-600 EFB

Paperless Cockpit GA-EFB 4000

Photos courtesy of Paperless Cockpit.

Note: Paperless Cockpit is working on the next generation of these devices and expects to announce them in October 2003.

Product Name	Paperless Cockpit LT P-600 EFB, Paperless Cockpit GA-EFB 4000
Website(s)	• Paperless Cockpit • LT P-600 EFB • GA-EFB 4000
Display Size	LT P-600 EFB: 8.4", 800x600 SVGA GA-EFB 4000: 10.4" TFT, XGA
Controls	Touch screen, indoor/outdoor viewable, active displays, wireless keyboards, and more options available
Mounting Style	Various mounting solutions including yoke mounts, articulating armatures, kneeboards, and custom docking stations
Form Factor	LT P-600 EFB: Fujitsu Stylistic LT P-600 GA-EFB 4000: Fujitsu ST 4000
Operating System	Microsoft Windows OS
Applications Supported	
Electronic Charts	✓ (Jeppesen JeppView FliteDeck)
Electronic Checklists	✓ (OBDS)
Electronic Documents	✓ (Adobe Acrobat)
Flight Performance Calculations	✓ (Ultra-Nav)
Flight Planning	✓ (FliteMap, FliteStar, RMS Tech FlightPlan)
Video Surveillance	✓
Other	Any Microsoft Windows compatible application. Weather applications include Echo Flight, WSI, and XM WX Satellite Weather's WxWorx on Wings for in-flight use or Jeppesen Weather or DUATS for flight planning. Also provides software for maintenance, reporting, scheduling, and custom applications and access to FOM, SOP, MEL, CDL and OEM Tech Pubs.
Approvals	
Potential Customer(s)	Class 1 or Class 2 EFB Air transport, business jet, high- and low-end GA, military

Appendix A   169

14. Teledyne Technologies (formerly Spirent Systems)

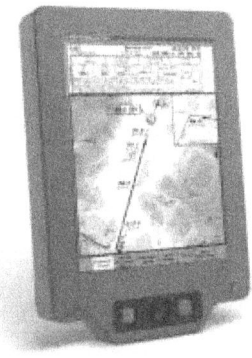

Product Name	AvVantage
Website(s)	• Teledyne • AvVantage • July 2002 Avionics Magazine article
Display Size	8.4" or 10.4" displays
Controls	6 soft function keys, touch screen
Mounting Style	Certified mount/dock
Form Factor	Tethered and integrated (one piece) versions
Operating System	Microsoft Windows operating system
Applications Supported	Open architecture; able to be networked
Electronic Charts	✓
Electronic Checklists	✓
Electronic Documents	✓
Flight Performance Calculations	✓
Flight Planning	
Video Surveillance	✓
Other	Fault reporting
Approvals	STC for AvVantage on A300
Potential Customer(s)	• Qantas Airways • FedEx [Totally Integrated Technical Aircraft Network (TITAN) System] • Evaluated by Continental • In service at 6+ operators on 7+ aircraft models • 70+ units in service • In cooperative development agreements with multiple potential customers (e.g., Continental, Embraer)

Photos courtesy of Teledyne Technologies.

15. Universal Avionics

Photo courtesy of Universal.

Product Name	Universal Cockpit Display (UCD) / Universal Cockpit Display Terminal (UCDT-II)
Website(s)	• Universal Avionics • UCD product information • July 2002 Avionics Magazine article
Display Size	UCD: 10" diagonal display, 780 x 1024 UCDT-II: 8.4" diagonal display, 800x600
Controls	Touch-screen
Mounting Style	Tethered; cockpit or yoke-mountable
Operating System	
Applications Supported	All software must be purchased from Universal
Electronic Charts	✓
Electronic Checklists	✓
Electronic Documents	
Flight Performance Calculations	
Flight Planning	
Video Surveillance	✓
Other	
Approvals	• Certified to DO-160 with Level C software (DO-178B) with TAWS TSO, and electronic approach charts (internally developed software, not Jeppesen) • UCDT-II approved for 10 business aircraft, including Falcon 10, 20, and 50, King Air 350, Boeing Business Jet, and Bombardier's Challenger and Global Express. In the process of certification for Cessna Citation and Gulfstream. Also certified for U.S. Air Force RC-135
Potential Customer(s)	Air transport and business jet

Appendix A   171

1.2 Software and Content Providers

The following table lists vendors who are developing software that for EFBs. The table below provides information on the application type (e.g., electronic charts, checklists, etc), the operating systems supported, the hardware platform on which the software can be used, and potential customers. Note that the provider and product names are hyperlinked to the corresponding websites, where possible. Their URLs are included at the end of this appendix.

Provider	Product Name	Electronic Charts	Electronic Checklists	Electronic Documents	Flight Perf. Calc.	Flight Planning	Video Surveillance	Other Applications	Operating Systems Supported	Hardware Platform	Customers
Adobe	Acrobat Reader			✓					Microsoft Windows, Pocket PC, Palm, and Symbian OSs		
Airbus	FCOM/OEB			✓							
	Less Paper Cockpit (LPC)				✓						
Aircraft Data Fusion	xEFB	✓	✓	✓			✓	Weather, TCAS, terrain, ACARS, surface management, eLearning		Astronautics PID	
Aircraft Management Technologies (AMT)	Flightman	✓	✓	✓	✓	✓	✓	Voyage report, maintenance logs, and on board sales/ cabin management. Communications Manager (WiFi, Satcom, ACARS, GSM/GPRS)	Browser-based application. Run on Microsoft Windows, Linux, or other UNIX OS	Any multi-functional display, portable computer (tablet, laptop, or PDA) Flightman and eTechLog can be configured to any aircraft type or hardware type (Class 1, 2, 3)	Pricing expected to vary based on custom features, from mid-range to high end • Atlas Airways • Futura • DHL
	eTechLog							Technical log system. Manages a/c defects and generates troubleshooting procedure and accurate defect/parts information	Integration to airline legacy systems via widely used industry standards (IBM MQ Series, Web Services)		
Astronautics	Pilot Information Display (PID)							CPDLC, ACARS, Video, CDTI	Linux OS	Astronautics PID	
Control Vision	Anywhere Map	✓	✓			✓				PocketPC	GA
Echo Flight	EchoMap							Moving map with data link weather information	Microsoft Windows 95/98/NT/ 2000/XP laptop		

Appendix A

Provider	Product Name	Electronic Charts	Electronic Checklists	Electronic Documents	Flight Perf. Calc.	Flight Planning	Video Surveillance	Other Applications	Operating Systems Supported	Hardware Platform	Customers
Flight Deck Resources	Chartrax	✓								SkyTab EFB	Part 91/121/135/125
FlyTimer	FT2000		✓		✓	✓			Palm OS	Palm	GA
Honeywell	WINN							Weather application	Microsoft Windows NT/2000/XP	Laptop/pen tablet	
ION Systems	eMonocle		✓						Microsoft Windows OS		
Jeppesen	JeppView FliteDeck	✓								Portable computer (tablet/laptop/PDA)	GA and business jet Universal UCD and Garmin AT use Jeppesen data, but not the Microsoft Windows software interface
	FliteMap					✓					
	FliteStar					✓					
	EFB (Class 1)	✓			✓	✓		Standard Developer's Kit (SDK) for 3rd party app devel.	Microsoft Windows 95/98/NT/2000/XP	Laptops, Pen Tablets	Part 91/121/135/125, Military
	EFB (Class 2)	✓			✓	✓	✓	Taxi Position Awareness moving map with ownship position, eLogbook, SDK for 3rd party applications in development, enroute moving map with Wx in development.	Microsoft Windows 95/98/NT/2000/XP	Laptops, Pen Tablets, custom devices	Part 91/121/135/125, Military
	EFB (Class 3)	✓			✓	✓	✓	Taxi Position Awareness moving map with ownship position, eLogbook, SDK for 3rd party applications in development, enroute moving map with Wx in development.	Microsoft Windows 2000 for all except Linux for TPA	Astronautics PID	Part 121/125, Military, high end Part 91/135

Appendix A   173

Provider	Product Name	Electronic Charts	Electronic Checklists	Electronic Documents	Flight Perf. Calc.	Flight Planning	Video Surveillance	Other Applications	Operating Systems Supported	Hardware Platform	Customers
Lido	Aircraft Performance Services				✓					PC or notebook	Air transport, business jet, high- and low-end GA
	Briefing										
	FlightNav	✓									
	Flight Watch					✓					
	Integrated Take-off Performance (LinTop)				✓						
	Take-off Performance Analysis (TOPAS)										
On Board Data Systems	Multi-Function Electronic Flight Bag (MF-EFB)	✓	✓	✓	✓		✓	Wireless Document Library Synchronization User defined GUI	Microsoft Windows OS	All Class 2 EFB and Network PC or Wireless Devices	Flight Crews Fleet Management Commuter Business Aviation
	Electronic Pilot's Advisory Data System (EPADs)	✓	✓	✓	✓		✓	User-Managed Doc-Lbrary for Fleet Synchronization/ Distribution/ Subscription ECL Specialists	Microsoft Windows OS Shell User-defined Multitasking GUI	Supports all Class 2 EFB, PCs, & wireless devices	Fleet ops, commuter, and corporate
RMS Technology	Flitesoft					✓			Microsoft Windows OS		
	Vista							Moving map			
Rockwell Collins Press release	Video Intelligence System (VIS)						✓				UAL
Stenbock & Everson	FlightPrep	✓				✓		Weather		Pocket PC	
	Polaris Moving Map							Moving map			Avrotec
Teledyne Technologies (formerly Spirent Systems)	Cargo Load Planning System (CLPS)				✓				Microsoft Windows operating system	Fujitsu, Walkabout Computers (Hammer-head model)	
	Flight Operations Documentation (FOD)			✓							
	Ground Weight & Balance System (GWBS)				✓						
	Onboard Performance System (OPS)				✓						
	Pilot Fault Reporting (PFR)										
	Realtime Graphical Weather (RGW)										
	Video Surveillance Application						✓				

Appendix A

Provider	Product Name	Electronic Charts	Electronic Checklists	Electronic Documents	Flight Perf. Calc.	Flight Planning	Video Surveillance	Other Applications	Operating Systems Supported	Hardware Platform	Customers
Ultra-Nav	Ultra-Nav				✓				Microsoft Windows and Palm OS		
WSI	InFlight							Weather		Fujitsu (P-600 or ST-4000)	
Weather WxWorx	XM WX Satellite Weather's WxWorx on Wings – Basic XM WX Satellite Weather's WxWorx on Wings – Premium							Weather	Microsoft Windows 2000, Microsoft Windows XP	Tablet PC	

Appendix A   175

1.3 Display Providers

The following is a list of vendors who develop displays for EFB systems. For each display, we note the website(s) where more information can be found, the display size, controls, mounting style, approval (if any), and potential customers. The website names listed here are hyperlinked. Their URLs are included at the end of this appendix.

1. Astronautics: See information in EFB Systems in Section 1.1 (3)

2. CMC Electronics (formerly Northstar): See information in Section 1.1 (7)

3. Fujitsu

© 2003 Fujitsu Computer Systems.
Permission to use, copy and distribute documents delivered from this World Wide Web server and related graphics is hereby granted, provided that the above copyright notice and this permission notice appear. All other rights reserved.

Product Name	Fujitsu Stylistic LTP-600
Website(s)	• Fujitsu • Stylistic LTP-600
Display Size	8.4" transreflective SVGA TFT
Controls	Programmable and dedicated bezel keys, touch screen, and external keyboard
Operating System	Microsoft Windows
Mounting Style	
Potential Customer(s)	• Hardware for content provided by ADR, ApproachView, Flight Deck Resources, Paperless Cockpit • Evaluated by Continental

4. HP/Compaq

Photos courtesy of Hewlett-Packard website.
© 2003 Hewlett-Packard Development Company, L.P.

Product Name	HP laptops and iPAQ
Website(s)	• HP • Laptops • Handheld devices
Display Size	Ranges in size from 3.5" (240x320) to 3.8" (240x320)
Controls	Touch screen display
Operating System	Microsoft Pocket PC
Mounting Style	
Potential Customer(s)	Air transport, business jet, high- and low-end GA

5. Panasonic

PDRC (12.1" x 1.2", 1024 x 768)

MDWD with Toughbook 07 (8.4", 800x600)

Copyright Panasonic Computer Solutions Company.

Product Name	Toughbook family
Website(s)	• Panasonic • Toughbook
Display Size	Depends on the model. Display sizes range from 3.4" 240x320 (QVGA) to 15" 1600 x 1200 (UXGA).
Controls	Depends on the model
Operating System	Microsoft Windows
Mounting Style	
Potential Customer(s)	• Supplier of rugged mobile computing solutions to the U.S. military. Toughbooks are the defacto standard for USAF, USMC, Navy and most any aviation application for the military and are used in aircraft ranging from B-52's to BlackHawks. • Also interested in air transport, business jet, high-end and low-end GA

6. Paperless Cockpit: See information in EFB Systems in Section 1.1 (13)

7. Teledyne Technologies: See information in EFB Systems in Section 1.1 (14)

8. WalkAbout Computers

Photo courtesy of Walkabout Comp.

Product Name	HammerHead 3 (HH3)
Website(s)	• WalkAbout Computers • HH3
Display Size	10.4", 800 x 600 active matrix TFT
Controls	Keyboard; custom interfaces available
Operating System	Microsoft Windows
Mounting Style	
Potential Customer(s)	Partnership with Teledyne Technologies to outfit all of GB Airway's Airbus 320/321 aircraft as Hardware Class 2 device

2 Military Activities

The US Department of Defense is showing interest in EFB technology. A military report providing an overview of EFBs was published in August 2002 (Fitzsimmons, 2002). The report discusses potential military benefits for EFBs and gives examples in the appendices of the advantages for an EFB for a resupply mission by cargo aircraft and combat mission in a tactical fighter. Specific military applications for EFBs include flight performance calculations for mission planning, real-time updates of intelligence briefings, and the presentation of real-time and near real-time information overlaid on images and charts.

Additionally, a military conference on EFB was held at the US Air Force Academy (USAFA) in March 2003. The conference was organized by the USAFA Institute for Information Technology Application (IITA). It included presentations by several military agencies describing their efforts to develop EFB hardware and applications. Several EFB hardware and software vendors also gave presentations and brought demonstration platforms for the audience to examine. The presentations consisted of the following:

- Pentagon—overview of Air Force EFB concept and status of initial requirements document
- AF Air Mobility Battlelab—Real Time in the Cockpit project description
- NIMA—Electronic information products (e.g., geo-spatial intelligence)
- Wright Patterson AFB—Electronic flight publications (documents and checklists), including human factors evaluations
- Air Force Research Lab—Digital Kneeboards for ejection seat aircraft

3 Industry References

- Fitzsimons, B. (2002, May). Jeppesen more than just paper. *Aviation International News.*
  Website: www.ainonline.com/Publications/EBACE/EBACE_02/ebace_02zzday3jepp html
- Fitzsimmons, F.S. (2002). The Electronic Flight Bag: A Multi-Function Tool For The Modern Cockpit. Institute for Information Technology Applications: United States Air Force Academy, Colorado.
  Website: www.usafa.af mil/iita/Publications/ElectronicFlightBag/EFBWebversion.doc
- Hanson, E. (2002, June). Electronic flight bags: What airlines want. *Avionics Magazine*, pp. 32-34.
  Website: www.aviationtoday.com/reports/avionics/previous/0602/0602efbs.htm
- Jensen, D. (2002, July) Electronic flight bags: An emerging market heats up. *Avionics Magazine*, pp. 35-42.
  Website: www.aviationtoday.com/reports/avionics/previous/0702/0702efb htm
- Jensen, D. (2001, April) Pilots, toss away your black flight bags. *Avionics Magazine*, pp. 39-40.
  Website: www.aviationtoday.com/reports/avionics/previous/0401/0401flightbag htm
- Ramsey, J.W. (2003, January). Eye on the Cabin. Avionics Magazine.
  Website: www.aviationtoday.com/reports/avionics/previous/0103/0103cabin.htm
- Ramsey, J.W. (2003, March). Weather Avoidance. Avionics Magazine.
  Website: www.aviationtoday.com/reports/avionics/previous/0303/0303weather htm
- Trotter-Cox, A. (March 2000) Electronic flight bag: Transitioning to a paperless environment requires more than technology. *Professional Pilot* (Volume 34 No. 3). Alexandria, VA. Queensmith Corporation.
  Website: www.aviationmanuals.com/articles/article6 html

Websites

The following is a list of websites for EFB system manufacturers, software and content providers, and display providers discussed in the industry review. Websites where more information can be found about a product are also included. This list was compiled in September, 2003. Please note that these links may become out of date fairly quickly.

1.1 EFB Systems

EFB Systems Manufacturer	Website
ADR	www.adrsoft.com
ApproachView	www.approachview.com
Astronautics	www.astronautics.com
Avrotec	www.avrotec.com
Boeing	www.boeing.com
	Information for Boeing customers: www.myboeingfleet.com
Jeppesen	www.jeppesen.com
C-Map/Aviation	aviation.c-map.com
CMC Electronics	www.cmcelectronics.ca/
	(formerly Northstar: www.northstarcmc.com/default.htm)
eflightsystems, LLC	www.eflightsystems.com
Flight Deck Resources	www.flightdeck.aero
GSCS	www.grid.com/products.html
JP Instruments	www.jpinstruments.com
NavAero	www.navaero.com
Paperless Cockpit	www.paperlesscockpit.com
Teledyne Technologies	www.teledyne.com
	(formerly Spirent Systems: www.spirent-systems.com)
Universal Avionics	www.universalavionics.com

1.2 Software and Content Providers

Software/Content Provider	Website
Adobe	www.adobe.com
Airbus	www.airbus.com
Aircraft Data Fusion	www.aircraftdatafusion.com
Aircraft Management Technologies (AMT)	www.airmantech.com
Astronautics	www.astronautics.com
Control Vision	www.controlvision.com
Echo Flight	www.echoflight.com

Appendix A

Software/Content Provider	Website
Flight Deck Resources	www.flightdeckresources.com
FlyTimer	www.flytimer.com
Honeywell	www.honeywell.com
ION Systems	www.ionsystems.com
Jeppesen	www.jeppesen.com
Lido	www.lhsystems.de/englisch/solutions/index.html
On Board Data Systems	www.obds.com
RMS Technology	www.rmstek.com
Rockwell Collins	www.rockwellcollins.com
Stenbock & Everson Flight Prep	www.flightprep.com
Teledyne Technologies	www.teledyne.com
Ultra-Nav	www.ultranav.com
WSI	www.wsi.com
WxWorx	www.wxworx.com

1.3 Display Providers

Hardware Provider	Website
Astronautics	www.astronautics.com
CMC Electronics	www.cmcelectronics.ca
Fujitsu	www.fujitsu.com
HP/Compaq	www.hp.com
Panasonic	www.panasonic.com
Paperless Cockpit	www.paperlesscockpit.com
WalkAbout Computers	www.walkabout-comp.com

Appendix B:
Summary of Equipment Requirements and Recommendations

These summary tables are provided as a quick reference for the equipment requirements and recommendations listed in the document. Notes that provide additional guidance on interpreting the requirements from the body of the document and issues pertaining to Installation or Training/Procedures and interactions between Equipment and Installation or Training/Procedures are not included here.

These summary tables are designed to be of use by manufacturers and FAA evaluators when conducting office-setting human factors evaluations of EFBs. The tables list the section and topic where more information can be found; these headings are cross-referenced with the corresponding sections in the document. Equipment requirements are designated with a ❖ and shaded. Equipment recommendations are designated using a ❑ and are not shaded. This is illustrated in the figure below.

Section numbers and topic headings listed and cross-referenced within the document.

Guidance presents a summary of the equipment requirements and recommendations.

Section	Topic		Guidance
2.1.1	Workload	❑	Flight crew workload and head-down time should be minimized (AC 120-76A, Section 10.c)
2.1.5	Legibility—Lighting Issues	❖	Automatic brightness adjustment should operate independently for each EFB
		❑	Screen brightness should be adjustable in fine increments or continuously
		❑	Buttons and labels should be adequately illuminated for night use

2 General EFB System

Section	Topic	Guidance
2.1.1	Workload	❏ Flight crew workload and head-down time should be minimized (AC 120-76A, Section 10.c)
2.1.5	Legibility—Lighting Issues	❖ Automatic brightness adjustment should be independent for each EFB (See AC 25-11) ❏ Screen brightness should adjustable in fine increments or continuously ❏ Buttons and labels should be adequately illuminated for night use
2.1.7	Failure Modes	❏ EFB should alert the flight crew to probable application/system failures (AC 120-76A, Section 10.e (2))
2.2.4	Kneeboard EFBs	❖ Kneeboard EFB should be easily removable
2.4.1	User Interface—General Design	❏ User interface should have a consistent set of controls and graphical elements ❏ Controls used for different purposes should be visually distinct from one another ❏ Graphic elements and controls should follow personal computer conventions, except where clearly inappropriate for flight deck environment ❏ Menu functions should be accessible in proportion to frequency of use and criticality to mission
2.4.2	Application Compatibility and Style Guides	❏ All applications should follow a common style guide, preferably specific to that aircraft ❏ Color and other formatting should be internally consistent across applications (AC 120-76A, Section 10.b (1)) ❏ Help facility, if available, should be standardized across applications ❏ Soft key labels and menus should be consistent across applications ❏ Common actions allowed on multiple applications should be performed in the same manner ❏ Manufacturers should prepare style guides for third party developers
2.4.3	General Use of Colors	❖ Red and amber should be reserved for highlighting *warning* and *caution* level conditions respectively (AC 120-76A, 10.d(1)) ❖ Color should not be sole means of coding important differences in information; color should be used redundantly ❖ Color-coding scheme should be interpretable easily and accurately. ❏ Each color should be associated with only one meaning ❏ No more than six colors with assigned meanings should be used in a color-coding scheme ❏ EFB colors should not conflict with flight deck conventions ❏ If colors are customizable, there should be an easy way to return to default settings
2.4.4	Graphical Icons	❏ Icons should be accompanied by brief text labels ❏ Design of icons should minimize training and maximize intuitiveness for cross-cultural use

Appendix B 183

Section	Topic	Guidance
2.4.5	Multi-Tasking	❏ The user should be able to identify the active application easily ❏ The user should be able to: - Select which of the open applications is currently active - Switch between applications easily ❏ Applications running in the background should be in the same state when the user returns to it, other than the completion of any background processing ❏ Responsiveness of an individual application should not suffer when all applications are running simultaneously ❏ The user should be able to exit applications with pending activities by completing them or by acknowledging that they are incomplete ❏ The system should discourage use of non-flight-related applications, and ask for an extra confirmation to launch
2.4.6	Responsiveness	❖ The system should provide feedback when a user input is processed - Alphanumeric inputs should be shown within 0.2 seconds (SAE ARP 4791) ❖ A "system busy" indicator should be displayed if user inputs can not be processed within 0.5 seconds (SAE ARP 4791) ❏ The EFB applications should have a "system busy" indicator ❏ The type of feedback should be appropriate for the type of user input ❏ If tasks take more than a few seconds to complete, indicators should show their progress ❏ User entries made while the system is busy should be stored for later processing
2.4.7	Anchor Locations	❏ If the EFB supports more than one application, there should be an anchor location from which the user moves between applications ❏ Each EFB application should have its own anchor page ❏ It should be easy to move from any location in the EFB to an anchor location, and vice versa
2.4.8	Alerts and Reminders	❖ Alerts and reminders should meet 14 CFR Part 23.1322, 25.1322, 27.1322 or 29.1322 as appropriate. Their intent should be generalized to the use of colors on displays and controls (AC 120-76A, 10.d (1)) ❖ Red should be used only for warnings (AC 120-76A, 10.d (1)) ❖ Amber should be used only for cautions (AC 120-76A, 10.d (1)) ❖ Other colors should be sufficiently distinct from red/amber for use (AC 120-76A, 10.d (1)) ❏ Alerts and reminders should be consistent with AC 25-11, 14 CFR Part 23.1311a, AMJ 25-11 ❏ Alerts should be integrated or compatible with other flight deck alerts (AC 120-76A, 10.d (1)) ❏ Messages should be prioritized and the prioritization scheme should be documented and evaluated (AC 120-76A, 10.d (1) and AC 120-76A, 10.d (2)) ❏ Strong attention-getting techniques (e.g., flashing or bright text) should be avoided (AC 120-76A, 10.d (1)) ❏ During high workload phases of flight: (a) Required flight information should be continuously displayed and unobscured, except when messages are needed to indicate failure or degradation of the EFB application (AC 120-76A, 10.d (1)) (b) Messages should be inhibited, except those that indicate failure or degradation of the EFB application (AC 120-76A, 10.d (1))

Section	Topic	Guidance
2.4.9	Display of System Status	❏ Any full or partial application failure should be indicated with a positive indicator (AC 120-76A, Section 10.d (2)) ❏ The immediacy of indicator should be appropriate to the function that is lost or disabled (AC 120-76A, Section 10.d (2))

2.4.10	Legibility of Text—Character	❏ Typeface should be highly legible. HFDS recommends:
		  - Spare use of upper case text (8.2.5.8.2)
		  - Mixed upper and lower case for continuous text (8.2.5.8.4)
		  - Serif fonts for high resolution displays (8.2.5.7.5)
		  - Sans serif fonts otherwise (8.2.5.7.6)
		  - Character contrast between 6:1 and 10:1 (8.2.5.6.12)
		  - Characters stroke width 10 to 12% of character height (8.2.5.6.14)
		❏ Individual characters should not be easily confused with other characters
		❏ Slanting or italic text should be avoided

2.4.11	Legibility of Text—Typeface Size and Width	❖ Typeface should be appropriate for viewing distance, lighting conditions, and text criticality
		❏ The FAA HFDS recommends that:
		  (a) Minimum character height should be 1/200 of viewing distance, e.g., for 35" viewing distance, 0.175" tall (17.5 pixels at 100 pix/inch) (8.2.5.6.6)
		  (b) Preferred character height should be 1/167 of viewing distance (8.2.5.6.5)
		  (c) Character height to width ratios should be (8.2.5.6.10)
		    o <80 char per line, 1 to 0.7 up to 0.9 (15 pix tall, 10.5 to 13.5 pix wide) for monotype fonts
		    o >80 char per line, at least 1 to 0.5 (15 pix tall, 7.5 pix wide)
		    o 1:1 for M and W in a proportional font
		❏ Larger fonts should be used for text read in poor viewing conditions

2.4.12	Legibility of Text—Spacing for Readability	❏ Text should be spaced appropriately to facilitate reading
		❏ Line lengths should be appropriate for text content
		❏ To facilitate readability, HFDS recommends the following:
		  (a) Use horizontal spacing between characters that is at least 10% of character height (15 pix tall, 1.5 pix spacing) (8.2.5.6.1)
		  (b) Use spacing between words of at least one character for equally spaced characters, or width of "N" for proportional fonts (8.2.5.6.2)
		  (c) Use spacing between lines of at least two stroke widths or 0.15 of character height (15 pix tall, 2.25 pix leading), whichever is greater (8.2.5.6.3)
		  (d) Separate paragraphs with blank line (8.2.5.6.4)

2.4.13	Non-Text Display Elements	❏ Non-text display elements should be distinguishable based on shape alone, without relying on secondary cues such as color or labels ❏ Non-text display elements should be designed for legibility on minimum expected display resolution viewed from the maximal intended viewing distance
2.4.14	Supplemental Audio	❏ Supplemental audio should be avoided in flight ❏ Users should be able to control the volume ❏ Users should be able to turn off the supplemental audio ❏ Objects with supplemental audio should be coded so the user knows of the associated audio before activating it ❏ Supplemental audio that is solely audio should have text description available ❏ Users should be able to stop the supplemental audio at any time

Section	Topic	Guidance
2.4.15	Ensuring Integrity of EFB Data	❖ EFB data should be checked prior to installation to ensure that they are accurate, current, and uncorrupted ❏ The EFB should check that the current date is within the valid date range ❏ The EFB should allow data with an effective date in the future to be installed ❏ The system should conduct a self-test to ensure that the data is current and generate a message to the flight crew if any data is out of date. The message should indicate where to go for further information.
2.4.17	Crew Confirmation of EFB Software/Database Approval	❖ The latest revision information should be available upon request
2.4.18	Links to Related Material	❏ A consistent philosophy should be used for accessing different types of information. Similar types of information should be accessed in the same way ❏ Users should be able to keep track of how to move between topics. Users should be able to return to the starting point easily
2.4.19	User-Interface Customization	❖ There should be an easy means to return all settings to their default values ❏ For Part 121 and 135, the default settings should be customizable only by an administrator ❏ For Part 91, the default settings should be specified by the manufacturer and configurable by the user
2.5.1	Pointing and Cursor Control Devices	❏ Input devices should be selected and customized based on the type and complexity of the entries to be made and flight deck environmental factors that affect its usability ❏ Performance parameters should be tailored for the intended application and for the flight deck environment ❏ Users should be able to rest and/or stabilize their hand when using the pointer or cursor control device ❏ Active areas should be sized to permit accurate selection with the pointer/cursor device under all operating conditions

Section	Topic	Guidance
2.5.2	Hardware Controls	❖ All controls should be properly labeled (14 CFR 23.1555, 25.1555, and 27.1555) ❖ All soft function keys should be properly labeled ❖ Inactive soft function keys should not be labeled or should use a visual convention to indicate that the function is not available ❑ Physical function keys should provide tactile feedback when pushed ❑ Key repeats should be filtered by the software if they occur too closely together ❑ Soft function keys should be drawn in a reserved space outside the main content area ❑ The same function should appear on the same function key, whenever possible ❑ Labels should be consistent ❑ Labels should be clear and brief ❑ Labels should use standard abbreviations; ambiguous abbreviations should be avoided ❑ Labels should be located near the controls they identify and should not be confusingly close to other labels or other controls ❑ Labels should be drawn in horizontal text ❑ Physical controls should generally be co-located with the display ❑ The most frequently used controls should be placed at the most accessible locations ❑ Controls should be designed to deter inadvertent activation
2.5.3	Display	❑ The physical nature of the display screen should minimize the likelihood that information will be obscured (e.g., from dust)

2.5.5	Keyboards	❑ Keyboard type should be appropriate for the given task
		    ▪ QWERTY type keyboards should be used for text entry
		    ▪ Numeric keypads are best suited for significant numeric entries
		❑ Keyboards should provide appropriate tactile feedback
		❑ Users should be able to rest/stabilize their hand to use the keyboard, especially during turbulence

3 Electronic Documents

Section	Topic	Guidance
3.2.1	Consistency of Information Structure	❖ The information structure of the electronic document should be consistent with that of the hard copy
3.3.1	Visual Layout and Structure	❑ Windows and frames should be placed and used consistently ❑ Sections of text should be separated with ample white space ❑ Data should be formatted into short segments, where possible
3.3.2	Minimum Display Area and Resolution	❑ The minimum document display area and resolution should be specified by the manufacturer ❑ Operators should meet the manufacturer-specified display area and resolution requirements for training and operational use

Section	Topic	Guidance
3.3.3	Off-Screen Text	❖ The existence of off-screen content should be indicated clearly and consistently (AC 120-76A, 10.b (7))
		❑ Whether or not it is acceptable for parts of the document to be off-screen should be based on the application and intended function (AC 120-76A, 10.b (7))
		❑ Information regarding the document length and the current place within the document should be continuously available
3.3.4	Active Regions	❖ Active regions should be clearly indicated (AC 120-76A, 10.b (8))
3.3.6	Figures	❖ The electronic version of a figure should show all the content of the paper version
		❖ The entire figure should be viewable at once, even if all the details are not readable
		❖ All the details should be readable, although the entire figure may not be visible when doing so
		❑ Figures should be displayed in their entirety with all details readable whenever possible
		❑ Text information should be provided for each figure, independent of whether the figure is shown in full, or marked by a placeholder
		❑ The user should be able to configure the figure for optimal viewing
		❑ If zooming is supported, discrete zoom levels should be available (e.g. view whole page) and the current zoom level should be displayed at all times
3.4.1	Moving to Specific Locations	❑ The cursor should be visible at all times (AC 120-76A, 10.b (7))
		❑ If links are supported: - Entries in the table of contents should be linked to its location in the text - Cross-references should be linked to each other within a document
		❑ Users should be able to return to the previous location in one step
3.4.2	Managing Multiple Open Documents	❖ The active document should be indicated continuously (AC 120-76A, 10.b (9))
		❖ The user should be able to choose the active open document
		❑ A master list of all open documents should be available
3.4.3	Searching	❑ Search functionality should be available
		❑ Users should be able to select the document(s) to include in the search
3.5.1	Printing	❖ Pages or sections selected for printing should be clearly indicated
		❖ The user should be able to interrupt and terminate printing immediately
		❑ Users should be able to select document subsets for printing
		❑ The printed document should have the same visual structure as the EFB electronic document
3.5.2	Animation	❖ Start/stop functionality should be provided. The user should be able to stop the animation at any time
		❖ Text describing the animation should be available even if the animation is not running
		❑ Animation should not be overused
		❑ If supplemental audio is provided, control of the audio and video should be integrated

4 Electronic Checklist Systems

Section	Topic	Guidance
4.2.1	Checklists Supported by the ECL System	❖ If normal checklists are supported, then *all* normal checklists should be supported ❖ If non-normal emergency checklists are supported, then *all* non-normal checklists should be supported ❖ Similar requirements apply for other checklist categories
		❏ The ECL system should indicate the location of unsupported (i.e., paper) checklists ❏ Non-normal checklists should retain as much commonality as possible with normal checklists
4.2.2	Information and Visual Layout/Structure of Electronic Checklists	❖ The resulting crew actions called for in the checklist should be identical for paper and electronic versions ❏ Layout of items should be similar to the paper version. Headings, sub-headings, and titles should be consistent (CAP 807) ❏ The format of the electronic checklist should make it clear which challenge is associated with which response (CAP 708)
4.3.1	Accessing Checklists	❖ All supported checklists should be accessible for reference/review at any time while the system is active
		❏ Normal checklists should be accessible in accordance with the normal sequence of use ❏ Electronic checklists should be as quick and accurate to access as paper checklists ❏ The ECL system should open checklists only upon crew input
4.3.2	Managing Checklists	❖ The title of each open checklist should be visible continuously ❖ The checklist title should be displayed above the items and be distinguished throughout the checklist ❖ If more than one checklist can be open at once, other checklists should be accessible without closing the displayed checklist ❖ If more than one checklist can be open, the user should be able to select which one is active ❖ If a checklist is a "child" of another checklist, the user should be able to select whether the parent or child is active
		❏ A placeholder should be used to indicate which item was active prior to leaving the checklist ❏ The crew should be able to reset the checklist with a simple input ❏ Parent-child checklists should be integrated into a single checklist ❏ If more than one checklist can be open at once, a master list of checklists should be available
4.3.3	Managing Non-Normal Checklists	❏ All checklists associated with on-going non-normal conditions that are sensed should be listed on one master list ❏ A master list should indicate the status of each checklist
4.3.4	Lengthy Checklists	❖ The user should be able to look ahead (e.g., page down) without changing the active item
		❏ Information regarding the length of the checklist, the user's current position within the checklist, and how much of the checklist has been completed should be continuously available ❏ It should not be possible to change the status of off-screen items ❏ If the active item is off-screen and the user makes an "item completed" entry, an error message should appear or the active item should be called into view

Appendix B 189

Section	Topic	Guidance
4.3.5	Closing or Completing a Checklist	❖ If item status is tracked and the user attempts to close an incomplete checklist, the system should provide an indication that the checklist is incomplete and present any deferred/incomplete items for review ❖ The user should be able to close incomplete checklists only after acknowledging the indication
		❑ If item status is tracked, a positive indication should be presented when the entire checklist, as well as each item, is completed ❑ The action for closing/completing a checklist should be distinct from the action for marking an item as complete
4.3.6	Closing All Checklists	❑ The ECL should allow a state where no checklists are open ❑ The system should give a positive indication that no checklists are open; a blank screen is not sufficient
4.4.1	Indicating the Active Item	❑ The ECL should track and indicate the active checklist item ❑ When returning to an incomplete checklist, the item active prior to the move should again be active
4.4.2	Displaying Item Status	❖ Item status, if available, should be clearly indicated.
4.4.3	Moving Between Items Within a Checklist	❖ The active-item pointer should be moved to the next item with a simple action ❖ Returning to a previous item should not change the status of any item
		❑ If the status of individual items are tracked, the user should be able to: (a) Move from uncompleted items, changing their status to deferred (b) Move to the next item automatically after completing an item ❑ The user should be able to quickly select one item after another; system processing should not induce delays
4.4.4	Specifying Completion of Item	❖ User actions to mark an item as complete should be simple ❖ Completed items should not be removed from the screen immediately. The crew should be able to review the item and undo their action, if necessary
		❑ If the system indicates active items: a) The next item in the list should become active when an item has been completed, unless the next item is on the next page. A separate action should be required to move to the next page b) Moving to the next item without completing the current item should require an input distinct from that of specifying the item as complete ❑ An *undo* function should be available ❑ The completion status of each checklist item should be indicated clearly
4.5.1	Links Between Checklist Items and Related Information	❑ The navigation between links in the ECL and related information should be simple and clear ❑ Related information should appear in a single window or area of the screen. Hyperlinks from the related information should be shown in the same window or area

Section	Topic	Guidance

4.5.2	Links to Calculated Values	☐ If the EFB provides calculation worksheets and allows integration between these worksheets and the ECL system, then:
		  (a) Direct access to the appropriate worksheet should be provided for all items that can be calculated. This should be available for initial calculations and subsequent review/modifications
		  (b) The user should be able to return easily to the checklist item from which the worksheet was accessed
		  (c) Calculated ECL values should appear in the corresponding checklist location. These fields should be blank prior to inserting the calculated value

4.5.3	Task Reminders	☐ Reminders for high priority, time-critical tasks should be displayed continuously once in progress and should attract attention when delayed actions should be performed ☐ If multiple task reminders can be shown, crews should be able to determine how many are in progress and to what tasks they refer
4.5.4	Checklist Branching	☐ The selected branch should be clearly indicated ☐ The user should be able to back up and select another decision branch ☐ Items not on the selected branch should not be selectable

5 Flight Performance Calculations

Section	Topic	Guidance
5.1.1	Default Values	☐ Blank data entry fields should be used to indicate that there is no system assigned default value
5.1.2	Data-entry Screening and Error Messages	☐ The EFB should not accept user-entered data that is of incorrect format or type. Error messages should point out suspect entries and specify the expected data type. (AC 120-76A, Section 10.d (3)) ☐ The system should detect input errors as early as possible during data entry (AC 120-76A, Section 10.d (3)) ☐ The system should only discard erroneous input errors and not the whole set of entries related to the task in progress ☐ The system should present an error message when required values are missing; this error message should contain the name of the required value, using the label from the input field
5.1.3	Support Information for Performance Data Entry	❖ The units of each variable should be clearly labeled ☐ Labels, formats, and units of variables should match that in other sources (e.g., paper reports, flight deck systems) ☐ Related information for cross-checking should be in view or easily accessible
5.1.5	Modifying Performance Calculations	☐ The user should be able to modify previously computed results quickly ☐ Output relevant to earlier calculations should be erased once the user begins modifying those calculations

Appendix B 191

6 Electronic Charts

Section	Topic	Guidance
6.2.1	Transition from Paper to Electronic Charts	❑ Information structure of electronic charts should match that of paper charts ❑ Visual structure of electronic charts should be compatible with paper charts
6.2.2	Updates to Electronic Charts	❑ Corrections/updates should be made directly within the electronic chart application, unless they are temporary ❑ Corrections/updates that are of high priority or time-sensitive should not be made via paper notifications
6.2.3	Hard Copy Backups of Electronic Charts	❑ If the hard copy is used as a backup, it should be of sufficient quality to be used as effectively as the original paper chart. In particular: (a) The hard copy should be legible; all chart details should be visible (b) The quality of the paper should be acceptable for normal use (c) Color information should be distinguishable in the monochrome hard copy (d) All the chart information should fit on one printed page (e) The hard copy should be at least as large as a standard paper chart (f) The user should be able to select the size of the hard copy
6.2.4	Scale Information	❖ Scale information should always be visible for charts drawn to scale ❖ Scale information should be accurate. Scale information should be updated when the display is zoomed ❖ Static scale information should be removed unless it is always accurate ❖ Charts drawn "not to scale" should have a label indicating that fact continuously
6.2.5	Basic Zooming and Panning	❑ If zooming is supported, then panning should also be supported, and vice versa ❑ The chart's visual edges should be clearly marked. Visual edges should be shown only when no more information is outside that area ❑ When panning, the user should know which way to move to bring more of the chart into view ❑ Panning to an area where no portion of the chart will be displayed should be prevented ❑ If the user can change zoom levels, the user should be able to return to a default view easily ❑ If the display can be panned, the user should be able to return to a default view easily ❑ Zooming and panning should not result in lengthy processing delays

Section	Topic	Guidance
6.2.7	Orientation of Electronic Charts	❖ Orientation of the charts should be indicated continuously ❖ When charts are oriented with respect to directionality (e.g., track/heading), and directionality information becomes unusable, it should be clear to the pilot that that information is not available
		❏ When charts are oriented with respect to directionality (e.g., track/heading), and directionality information becomes unusable, (a) The crew should be notified of the unusable directionality and informed that the charts must revert to north-up orientation. (b) After crew acknowledgement of the failure, the charts should revert to the north-up orientation, the chart orientation indicator should be updated, and any cues that could imply directionality should be removed ❏ Text and symbols other than those designed to reflect compass orientation should remain upright at all times ❏ Crew input should be required to change the orientation of the charts
6.2.9	Access to Individual Charts	❖ The currently selected chart's label should be displayed continuously ❏ The system should allow rapid access to pre-selected charts ❏ The chart application should help the crew ensure that the correct chart was selected and allow corrections to be made quickly when an error occurs ❏ Multiple search methods should be supported ❏ Search results should be ordered with the most likely selections at the top of the list and least likely selections at the bottom ❏ Selection of alternate runways should be facilitated during approach
6.2.10	Knowledge and Display of Own-Aircraft Position	❖ Display of ownship should not be supported on non-georeferenced or not-to-scale terminal charts ❖ See TSO C-165 and DO-257A for other applicable requirements ❏ The range of display zoom levels should be compatible with the position accuracy of the ownship symbol. ❏ An indication of ownship position should be provided if the chart is zoomed or panned such that ownship is not in the current view
6.2.11	De-cluttering and Display Configuration	❏ The pilot should not be able to declutter safety critical display elements without knowing they are suppressed ❏ Changing map scale, orientation, and other options and settings should not induce significant levels of workload ❏ The information prioritization scheme should be documented

www.ingramcontent.com/pod-product-compliance
Lightning Source LLC
Chambersburg PA
CBHW071757200526
45167CB00017B/340